JOAN BAKER's background as a business financial consultant, in senior management, marketing and human resources, has been invaluable in giving her a deep appreciation of the opportunities and challenges facing today's farming community. Living as she does in the heart of Central Otago, Joan has advised many in the rural sector as they prepare for retirement. She is adept at managing successful change at both an individual level and in the wider context.

www.wealthcoaches.net
jbaker@wealthcoaches.net

Also by Joan Baker:

Improve Your FQ (Financial Quotient), 2003

Get Rich, Stay Rich (with Martin Hawes), 2003

Get Set By 30 (with Annette Sampson), 2004

Smart Women, Smart Money, 2004

Coach Yourself to Wealth, 2005

Live the Dream, 2006

A Man is Not a Financial Plan, 2007

Your Last Fencepost

S̲u̲ccession and R̲e̲tirement P̲l̲anning *for* N̲e̲w Z̲e̲aland F̲a̲rmers

Joan Baker

SHOAL BAY

Acknowledgements:
I would like to acknowledge the invaluable discussions with and feedback received from:
Sue Cumberworth, The Agribusiness Group, Christchurch
Ian and Gloria Hurst, Hurst Holdings, Oamaru
Malcolm Johnson, ex business risk adviser, Invercargill
Fraser McKenzie, AWS Legal, Invercargill
Desirée Reid, farmer, Timaru
Lynaire Ryan, Dexcel, Hamilton
Bill Thomson, Shand Thomson, Balclutha

This book is copyright. Apart from any fair dealing for the purpose of private study, research, criticism or review, as permitted under the Copyright Act, no part may be reproduced by any process without the written permission of Longacre Press and the author.

Joan Baker asserts her moral right to be identified as the author of this work.

© Joan Baker
ISBN 978 1 877361 21 0

A Shoal Bay book first published by Longacre Press, 2008
30 Moray Place, Dunedin, New Zealand

A catalogue record for this book is available from the National Library of New Zealand.

Book design by Book Design, Christchurch
Cover design by Christine Buess
Front cover photograph by davidwallphoto.com

Printed by Astra Print, Wellington

www.longacre.co.nz

Contents

Introduction: *Your last fencepost* 7

1. What's stopping you? 13
2. Communicate! 20
3. Treating siblings fairly 26
4. Family meetings 34
5. Identifying and preparing your successor 46
6. Getting ready for change 60
7. The numbers game 71
8. Investing off-farm 80
9. Reallocating your wealth 91
10. Income after your last fencepost 98
11. Wills, trusts and succession 104
12. Prescription or discretion? 109
13. Family trust basics 117
14. Being prescriptive 127
15. Getting good advice 132
16. After your last fencepost 142

Introduction:
Your last fencepost

I believe that farmers should have wonderful lives – all of their lives, even after leaving the farm. They have worked hard for decades to grow wealth and should therefore make sure that they will have what they need and desire in later life. Whether they wish to sell the farm to outsiders or pass it on to the next generation, the planning needs to start early in order to get the best outcomes for all. *What* you want is largely philosophical and each family has to work out what that looks like. *How* to get what you want, however, means that you have to understand quite a bit about succession, investment, asset allocation, family trusts, estate planning and so on. However, once you know what you want it is relatively easy to make it happen – especially if you begin early enough.

Many who farm land love the land they farm. However, farming is increasingly seen as a business as well as a way of life. Professional disciplines are being applied to all aspects of farming – strategy, operations, profit management and the creation of shareholder wealth. Succession too is one of the areas where a professional approach has much to offer. Family farms differ from corporate businesses in many respects but can still borrow good succession-planning concepts and processes from corporates. My experience of best practice by corporates is that they make planning for succession a core activity. They focus on management succession and continuity while farmers often focus solely on passing on ownership of the farm.

Believe it or not, there will be a last fencepost! Over a lifetime of farming you will put in a lot of fenceposts. It seems like a task that is never completed – as soon as you have got around the farm once, it's time to start at the beginning again. But there is an end – sooner or later you will finish your life as an active farmer, and will be putting in your last fencepost. On that day you will want to feel that you have had a good life in farming. You will also want to know that you have made the right decisions for your family about the farm and about the life you will lead in the future. This book is all about getting you happily to that last fencepost.

I come from farming families on both sides. We all care deeply about each and every *field* – just as for New Zealanders, the emotional attachment of Irish people to the land is legendary. Both my mother and father come from farms and, happily for us, both of these farms are still in the family. Though I live in New Zealand I am never happier than when stomping across those fields where I belong. However, both in my own family and in the wealth consultancy work I do, I have been astounded by the lack of planning around farm succession and the second half of farmers' lives. Many rural families experience great stress as the time approaches for the farm to be sold or passed on. But it is common for the issues not to be spoken of at all! Little wonder that so many farming couples do not get what they want – or even what they need – for their remaining decades. In addition there is often an unfortunate legacy of bitterness on all sides – dreams unfulfilled, schism in the family, unhappy siblings and loss of wealth.

This trauma is avoidable.

One day the farm will pass to someone else. You too will 'succeed' to a new phase of life: that's a given. For you to achieve the transfer on your terms, you need to have a succession planning process in place that will ensure that the land, business, knowledge and traditions are passed on successfully to the next generation and that you get to enjoy the new stage of life of your choice. So you have several questions to address:

- When will succession happen?
- Who will get the farm?
- How will you treat all of your children fairly?
- What money will you have?
- How will you avoid crippling your successor/farm with debt?
- How will you communicate regarding succession?
- How will you transfer the management of the farm?
- How will you transfer the ownership of the farm?
- Will you plan and control the process or just see what happens?

The last question is the most critical of all. And remember that you are planning *for the rest of your life.* Life expectancies have risen so much over the last generation that those who are looking to ease back from farming today may still have twenty, thirty or forty years of active living to plan for. What do you want? How much do you need? Will you have enough? How will you plan for an income? Where will you live? What will you do with all of your talents and time? These are very big questions to address and it is essential they get plenty of consideration. You need to start as early as possible to make sure that everything is in place to give you the outcomes and succession you desire. You also need to be mindful that succession may happen at any time – unexpectedly – and procrastination can be very costly on several fronts.

Recent research in New Zealand indicates that over 20 per cent of farmers are ready to retire or scale down their commitment to the farm; however, 40 per cent of these farmers have no succession plan at all. This is in line with what I see in my own work. Retirement or passing on of the farm is not a single event: it's a *process*. You have to decide what would suit you and your family and plan the process so that you can cut back or get out altogether on your own terms. To get the future that you want you have to create it – and it takes time. The numbers matter – the value of the farm, the amount of capital you need, the wealth you might transfer – but it's also about much, much more than just the finances.

Farm succession can be difficult. After all you may be owners, managers, trainers, workers, parents, trustees and often grandparents and so have multiple roles that can be in conflict when you come to consider farm succession. What you may desire as a parent may conflict with your needs as an owner or your good sense as a manager. Farms are very busy places too and it is often very difficult to get your head out of the daily grind – to stop looking at your gumboots – for long enough to consider the much larger and longer-term issues. There is usually only one asset – albeit a very big asset – the farm itself. This can make succession more complicated, especially where there is more than one child.

I have seen some glorious outcomes – and heard some gloomy stories. The difference is largely in the attitudes and behaviour of the farming couple. Will they think and talk about what they want? Will they discuss their hopes and expectations with the family? Do they solicit the hopes and dreams of the next generation? Starting these conversations as early as possible is important – there should be no surprises! Having lots of conversations over the years and regular family meetings helps to smooth the way. Starting these conversations is often the biggest hurdle for families in my experience and that's understandable. Our culture still makes topics like death, inheritance and personal money largely taboo; however, they must be addressed if succession is to work well. The best farm successions are planned and executed over a long time so that the transition from one generation to the next is seamless and feels completely natural for everyone involved.

All of this can seem quite daunting – it need not be. Changing from one way of life to another can be difficult even when we are getting what we want. Often it is not the change itself that is difficult but rather the *process of changing.* There are new and sometimes difficult conversations to be had and decisions to be made. Several chapters of this book deal with the issues around getting ready for a change and planning for a new life. You will also have to address all of the family issues such as choosing a successor and treating other siblings fairly. Making the best use of the advisers available to you is essential – to make your transition work smoothly and give you the outcomes that you want.

Money matters too. Many farmers struggle with the affordability issues around succession. They worry about whether they can afford to give up full-time farming and how to arrange their finances so that there will be enough income for them in retirement without overburdening the farm (and younger farmer) with a crippling debt. Both generations need to have enough income. I have devoted considerable space in this book to helping you work out how much you will need, how to invest off-farm and how to get income from your investments.

My experience is that it is the 'softer' aspects of succession that are most neglected – the thinking, the talking and the planning. However, the 'hard' or technical aspects matter too. It is very important to set up the right structures to facilitate succession and to organise your assets so that they can be easily and efficiently transferred. Much of this book is devoted to these technical aspects of succession. However, once you are clear about what you want your lawyer, accountant and other advisers can usually put the technical aspects together – as long as you give them the timeframes required.

It's time to stop looking *down* towards your gumboots and instead to look *forward* and plan the future life you want and deserve. Getting succession right – getting to the last fencepost effectively and harmoniously – is the ultimate challenge of the farmer's life. Identifying where you are at and what's holding you back is a good place to begin.

Chapter 1

What's stopping you?

Getting started

Sometimes the hardest part of any big project is getting started. There can be several reasons why we are reluctant to begin. It can really help to identify how you are feeling and what the blockages to progress are. It's particularly important that the senior farming couple are able to talk about succession issues as freely and frankly as possible, including how they are feeling about the various issues. Many of the farmers I meet delay thinking about succession, easing back or retirement for one or more of the following reasons.

This is who I am!
Many people wonder who they will be once they no longer run the farm; so much of their identity is tied up in what they do, they can be understandably reluctant to let those roles go. Many farmers wonder if they will go from being the owner of a big farm – 'the big

farmer' – to just 'the old man'. The remedy here is to start to think about what you will do, and who you will choose to be, post full-time farming. This is all about planning for the second half.

It will make me seem old
Most of us associate stepping back or stepping down with the admission of age. We are conditioned to see 'retirement' as the end – or very near it – and to be avoided if at all possible. Our parents' generation never seemed to live long once they gave up work and we – consciously or unconsciously – want to avoid this perceived hazard. Our world is very youth focused and most of us don't want to be put into the category 'old' or 'retired'. However, deciding to do other things, to travel more, to direct rather than operate the farm, does not make you old in fact or appearance. Many farmers will live for several decades after leaving farming; this is not about age, but rather about planning for all of the things you probably couldn't do while farming full-time.

What will they think?
Many farming folk are concerned about what their family and friends will think. Farmers tend to belong to close-knit communities so everyone will notice and know about the changes. Will this alter your relationships with all these people? Will they lose respect for you? Will they think you are now old and 'past it'? You may not care at all but many people do. Think through what you feel about this and get your 'sound bites' ready; if you care what others think then it's always a good idea to have decided in advance what you will tell them. You will feel much more comfortable with any changes you are making if you have worked out a script for when you are questioned.

It will make the problems 'real'
Most of us deal with difficult issues or unpleasant tasks by procrastinating or ignoring them altogether. The issues around

succession often look tangled – there are so many points of view to accommodate, so much planning to do, so many things to decide. It's understandable that it's often the path of least resistance – pretending that none of these issues exist – that is followed. Farmers may feel that once they open up the succession question all of these issues will become 'real', and will have to be grappled with immediately and decided at once. However, the reality is rather different. Usually, the whole family (and everyone else who is affected) breathes a sigh of relief. And everything doesn't need to be resolved at once – in fact, it's a step-by-step process as I will show.

I might be bored
When you have spent all of your life farming – and farmers spend all their time at work in a way that few others do – it can be difficult to imagine what you will do, how you will spend your time and where you will get such an outlet for your energies. Your work is your vocation: you live at your work, your friends and community are based around the farm. Again, the solution is to think about all the other things you would like to do, all the activities you have never had time for or have put on hold for decades as you farmed, and raised a family. Think back to the things that interested you when you were much younger – this is often a good guide to true interests and passions.

What next?
The biggest block comes from not having a clear picture of what you will do, where you will live, how you will spend your time, how you will still make a contribution. When you are used to having every day accounted for – no matter how much you may have disliked that at times – it can be hard to face a blank canvas. Knowing what you *don't* want is usually easy – many farmers find it very easy to say that they never again want to be crutching or in the milking shed morning and night – but it is much harder to identify what you *do* want for the coming years.

What will we do for money?
Worrying about money is common even when there are valuable assets like a farm. Farmers are used to living frugally and many have lived off the farm as well, as regards provision of vehicles, fuel, meat and vegetables. It's not easy in my experience to convince even the wealthiest people that they have enough wealth. When you have spent your whole life trying to amass wealth and keep your business above water it can seem very strange to consider using that wealth in order to live – it goes against your most ingrained habits. Farmers are often worried about taking too much out of the farm, if they are passing it on to the next generation, and leaving a crippling debt. There will also be issues around other children: how can they be treated fairly without dividing up the farm or selling it off? A great deal of what follows in this book is aimed at helping you resolve these mental blocks.

Before you read on, it's a good idea to identify where you're at now in the ongoing process of getting to your last fencepost.

Where are you at?

- Do you know what you want?
- Are you ready to make a change?
- Have you a timeframe in mind?
- Do you have an exit plan?
- Do you have a retirement plan?
- Do you know what your spouse wants?
- Do you know what your children want?
- Have you discussed this as a family?
- Is there a plan in place?
- Is there a written plan?
- Have you done the numbers?

You may not yet be clear about what you want. If so, this is the place to start. You and your spouse need to begin the conversation about what you would like to do and have and be in the future.

You may know what you want and may have discussed it and agreed on it as a couple but not talked about it with your children. In that case, you will want to start the conversation with your children. There is a lot of guidance provided in chapter 4, 'Family meetings', though of course you can have these discussions with individuals at any time or place – over a meal, out on the farm, as part of your normal interactions.

Even if you have agreed on what you want and your children have been informed and involved in that decision you may still have work to do regarding the financing of the transition to your next stage. I think it is best to get all of these plans on paper and to formalise the timeframe and all the other important details. There is a lot of work to do here and many others may need to be involved such as your lawyer, accountant, banker and financial adviser.

I think it's really important when you are considering succession issues that you clarify your values and objectives regarding the process. Farming businesses are similar to other family businesses in that there is both a business and a family to consider. No matter how important the family is to you, you must think about the business too; no matter how businesslike you are and how successful the business is, you will have family issues to consider. This is always a juggling act for family businesses and it takes some skill to keep them in balance and to weigh up all of the issues involved when making a decision. One thing that will help you is to be mindful of the aspects of succession that really matter to you. Ask yourself which objectives apply to you:

- Keeping the farm in the family?
- Passing on a thriving business?
- Developing sufficient wealth for retirement?

- Ensuring a smooth transition?
- Preparing a successor?
- Treating all of your children fairly?
- Organising your wealth to provide income?
- Choosing your future lifestyle?
- Protecting the farm?
- Other objectives?

It is also important to be fully aware of how you are feeling. Emotion plays a very large role in our decision-making – and in our action or inaction. Be sure that you are as aware as possible of how are you feeling. It is helpful to name your feelings and identify what the feelings relate to. Are you:

- Excited about
  - opportunities for the future
  - getting to do other things off-farm
  - being free of the round-the-clock commitment?
- Engaged regarding
  - planning for succession
  - planning your lives for the future
  - the life you and your partner will succeed to?
- Anxious about
  - how succession will be resolved
  - having enough money for the future
  - how your partner will adapt to the changes?
- Worried that
  - there will be disputes in the family
  - your partner will not know how to keep busy after the farm
  - your farm won't sell for enough money
  - your successor will not do a good job?

- Afraid of
  - the future
  - the unknown
  - leaving the only life you know
  - family fallout?
- Conflicted about
  - going or staying
  - now or later
  - which child should succeed
  - what your partner wants?
- Concerned about
  - your partner's reaction
  - what the future holds
  - your successor's competence
  - your children's relationships?
- Feeling other emotions?

When you have established where you and your partner are at, and when you know what you want and are clear about what's important to both of you, then it's time to start thinking about communicating with the whole family.

Chapter 2

Communicate!

Communication is at the heart of a successful family business. That's a really easy thing to say but it is much harder to honour it in practice. Communication is the core of any relationship, business or personal. It's what people most often complain is lacking when there is conflict or a failure in a relationship. They never say there is too much communication; they almost always complain that there is too little. So what does this mean for a farming family?

My observation is that the happier and more successful farming businesses set up very good communication processes as early as possible. They involve as many of the family as they can and they include them in discussions about the farm and the family business from an early age. The farming business and the conversation about it surrounds them from the start – and the chat progresses from the high chair to the kitchen table to more formal settings.

Have lots of meetings

What this means in practice is that the family has regular meetings (see chapter 4) about the farm and its business, and all of the children are included. Obviously when children are younger much of this discussion is beyond them and they may only be included for part of the meeting. However, a real effort is made to help them understand that these meetings are important and that what is happening is relevant to their lives and their future. At the very least they learn that managing the business is important, that time is set aside for communication and that everyone can, and should, know what is going on and be included in decision-making.

Involve outsiders

It is also important that children meet and know the outsiders who are involved in the family business – the accountant, lawyer, farm adviser. These people are likely to attend some, if not all, family meetings. It is equally important that these outside advisers know and appreciate the children of the family as it is more than likely that they will be working with, and for, these younger members some time in the future. Outside advisers are in a great position to help with advice regarding the business education of these young people – some of whom are likely to end up owning the family business and others who will undertake other business and professional careers.

A family meeting is a great forum for passing on the traditions and heritage of a family and the land that it owns. A family meeting provides the ideal occasion to review past events, tell stories about the activities and achievements of the previous generations and instil a sense of stewardship and responsibility in the next generation. All of this is very important to a family that wants the land to remain within the family. However, it's a lifelong task: you cannot develop this sense of belonging and heritage at the last minute when it is time to hand on the farm.

The meetings also provide an opportunity to teach younger people as they become able about the business side of running a farm. It is relatively simple in these forums to help children understand the workings of the farm and how they translate into income and profit and wealth. Over time, children will understand the implications of decisions made and also witness the effects of unforeseen events on the fortunes of the farm. This is a great preparation for any path in life, whether or not a particular child ends up in farming.

Manage expectations

Set expectations from the start. As soon as children are old enough it is important that they know that the family farm is a business and that it supports their way of life. They can be helped to understand that they are participants in this farm. Children will have chores and will help out in various ways as appropriate for their age. As more is expected of them, so too should more information and inclusion be provided. It's about creating a sense that the whole family is in this together and that the fortunes of the farm affect them all.

Family meetings are also a great way to keep tabs on the developing attributes and competencies of each child. Children can be given roles at family meetings: introducing people, delivering brief reports, explaining how various activities work. As they develop they can present financials, explain the assumptions behind forecasts and participate in planning and decision-making.

Meetings are also a great place to explore the children's expectations. As long as they are asked and their contributions are welcomed children are usually quite forthcoming about sharing where their interests lie and talking about their hopes and expectations for the future. Some children will naturally have more interest in farming than others. Some will show an early fascination with machinery, animals, growing things or science, whereas other will display ease or comfort with organising things, numbers, ideas, language or communicating – possibly indicating that they are more likely to want to enter a career in law, accounting, management or business. That's to be expected. It's important that parents honour and support their children's growing interests and strengths – all the time being mindful that these are likely to change from year to year. However as time passes it becomes more and more obvious where an individual's ability and desire should lead – and it generally makes sense to advise young people to do what they want to do anyway!

Parents can use the opportunity afforded by regular family meetings to manage their children's expectations. For example, these meetings are a great place to discuss post-school education. It is important to encourage your children to become as skilled as possible while at the same time being clear about what you can afford to fund and support. Sooner or later there will be tradeoffs. The family may decide, for example, to support one child in achieving post-graduate qualifications in a professional area because it is agreed that this child will not succeed to the farming business. The family might agree to support another child in developing an off-farm business or in buying a farm elsewhere. I think it is important for family harmony and a sense of fairness that these discussions take place openly. It is also important that these decisions are formally noted and that there is a record of what each child has received in the way of monetary support. All too often this earlier help is 'forgotten' – it can save a great deal of grievance and argument later if everyone is conscious and aware of how others have been supported along the way.

Clarify the 'rules'

Family meetings provide an obvious place for the farming couple to define the 'rules of the game' for all of the family. I have seen many farming families where Dad, in particular, had ideas and expectations about the future that he never shared with anyone (sometimes not even his wife!) until it was almost time to go. This can have disastrous consequences. It's understandable in some ways: farming can be a very solitary business and consequently many farmers operate as loners. This can be useful when coming to terms with spending most of your day alone with stock or crops or the fencing, but it's a bad recipe for preparing for succession. Family meetings provide a great place for the farming couple to outline their hopes and expectations early on. It's fine and even desirable for the owners to signal that they would like one or more of their children to take over the farm. The owners might also signal some kind of timeframe for that – becoming more precise as children grow more competent and the family gets closer to transition time. It's fine for the owners to say what they do and do not want – after all, it is their farm. Parents can articulate over time what they would like to do, how long they wish to be involved in the farm – and in what way – and what they think is fair and reasonable in terms of supporting all of their children.

It's much better – and much easier – to start these conversations earlier rather than later in family life. Everybody knows where they stand and members of the younger generation do not have to spend their lives 'second-guessing' the thoughts of their parents. When this works well everyone knows what is going to happen in advance and can adapt. When no one knows what is going to happen there is often lots of grief, broken relationships and missed opportunities. Maturing sons and daughters will be tempted to flee the farm and make a life for themselves elsewhere if there is no resolution in sight.

These family meetings need to happen frequently. Every family is different but from both a business and family point of view it seems important that you meet quarterly or biannually at least. For some of the meetings it would be an advantage to have others present. You might want to involve trustees from your family trust or at least inform them of the purpose and progress of these meetings. This is a good idea as it also formalises the concept of the family meeting. As before, children may be present for some or all of each meeting depending on their ages and stages. These meetings should also take place on the farm – it's much better if your advisers visit you rather than the reverse and it underlines the fact that the meeting is about this business and this family.

Remember that the purpose of these meetings is to discuss key issues, set goals that everyone can share and create a culture of open communication where everyone affected knows what is happening and what to expect.

It's important that these meetings are led and facilitated well. This could be a good role for an outside adviser. You may also want increasingly to give your children roles at each meeting. While this can slow things down and be a little awkward at first it will pay dividends over the longer term in building a sense of involvement and responsibility.

Chapter 3

Treating siblings fairly

What's 'fair'?

Family dynamics have changed over the last generation or so. It is no longer an automatic decision to leave the farm to one son to the exclusion of the rest of the family. All siblings must be considered and it is important not only to be fair but to be seen to be even-handed in your treatment of each child. But how do you do this? The farm is not usually large enough to split and shared management/ownership structures have a poor history of success as they tend to lead to a great deal of conflict.

It is rare for all of the children to want an active involvement in the farm. The interests of those who leave to follow other careers or partners are clearly different from those who remain actively involved in the farm. I think it's very important not to end up pitting their interests against each other as almost no course of action regarding the farm will suit them all. After all, if you don't actively farm your only interest is in getting income from the farm as rents or dividends, whereas your primary focus is likely to be reinvestment and wealth creation if you are actively farming. Farms don't flourish when they are being fought over in this way! I see this problem often not only with farms but with other family businesses.

'Fair' does not imply 'equal'. The real issue for most farming couples is how to treat their children equitably or justly over time. How will you arrange to have enough wealth in order to provide fairly for other children without having to compromise the farm or your own post-farming lifestyle? It is important to begin as early as possible so that you have the maximum time to ensure equitable outcomes. There are several things you can consider to achieve this goal – this chapter discusses some options I have used.

Sell the farm

There are several reasons why you might consider selling up including having no children who are interested in the farm or feeling that there is no way to organise a fair distribution of wealth that will work in your circumstances. Selling up provides you with a lump sum that should ensure that you have plenty to fund the rest of your lives. You can choose to make distributions to your children at the time of sale or leave them bequests through your will. Remember that it's OK to sell – it's your farm and your life's work that has made it whatever it is. Selling up is also a good option if you feel that you have left your planning for succession too late. There will always be other options once you are cashed up; for example, you can use some of the money to buy another farm or to help your children get started in some other way.

Partnerships

You (and your family) may consider that the farm is a no-sell property – a heritage that you all want to preserve. However you may have more than one child who wants to farm, which can be a recipe for family discord. What can be done? Children can go into partnership to run and/or own the farm if the property is large enough to allow it. There can be issues with such an arrangement unless the children work well together – and unless their partners can maintain productive and harmonious relationships. If the farm is large enough you could subdivide at the start. Alternatively, the partnership can acquire more land over time and eventually organise the divvy up later.

Buy out

The problem with having most of the wealth tied up in one asset is that it can be difficult to release capital for you or any of the siblings who do not succeed to the farm. A possible solution to this is that the successor gears up the farm initially in order to provide you with a lump sum to fund whatever lifestyle you are choosing post full-time farming. Many farmers who have worked all of their lives to reduce debt are reluctant to borrow more but it makes lots of sense. Your successor, who is getting a farming business and at least some share of the land itself, is off to a good start and should be able to service this debt. Inflation largely takes care of the problem. The risk with this strategy is that there may be a downturn in the fortunes of farming or interest rates may skyrocket. However, these are reasonable risks of being in business and ones that younger people should be able to shoulder with some optimism. Over time the same process can be used to progressively buy out siblings if necessary.

Off-farm assets

While many farms are not brilliant at producing income they concentrate capital very well over a lifetime. The farm has been the wealth-creation asset in your life. However farming is very risky – almost anything that goes wrong around the globe can affect your farm and your farm wealth. Farmers are at the mercy of international events, such as movements in markets, trade rounds, interruptions to shipping because of terrorism threats and currency fluctuations. That's even before you take account of what can happen closer to home – regulations, political changes, weather events, changing tastes, environmental issues. You only have to ask yourself what would happen to your wealth base if New Zealand got foot and mouth disease next week to realise how fragile a farm-based wealth portfolio can be. So I am very keen on farmers progressively investing off-farm in order to broaden the base of their wealth and make it more secure.

Off-farm assets might include a share portfolio, property, bonds, managed funds, superannuation policies or whole-of-life insurances. Having off-farm assets is also very useful when it comes to helping other children. You can afford to leave the farm intact but use some of these assets to make a distribution to other children at appropriate times. While these assets may not be the equivalent of the capital value of the farm, they go a long way towards allaying accusations that one child received all and the others got nothing. I think there are very good reasons for you to build a portfolio of off-farm assets anyway, quite apart from succession. You will find these very useful for funding your retirement: many a farm has had its operations severely curtailed by trying to fund the retirement of the older couple as well as to meet all of the demands of developing the farm and supporting the younger family. Most of these assets are unlikely to transfer until your estate is settled but you can enjoy the security of knowing that you have a store of wealth to be distributed.

Farmers often choose to help a child get into another farm nearby by using their farm value as collateral. This allows the child to build up expertise in farming so as to meet many of the succession objectives. It also provides a natural 'hedge' against farm values: if farms appreciate greatly in value over the ensuing years your child has a valuable asset to sell in order to buy your farm from you and allow you to retire.

Family trusts

A family trust can help with fairness problems (see chapter 11, 'Wills, trusts and succession'). The family trust can own the land and buildings, and one of the children can lease the farm from the trust and farm it. This child will own the farming business but the whole family will continue to own the farm. This arrangement has the added benefit of surviving you, and distributions can be made to any beneficiary at any time. The arrangement can work in many ways:

- One child earns a living from the farm or is paid a farm-management fee and the others get some money as rent.
- Other children may receive distributions from the farm over time as profitability allows.
- The agreement may be that the farm will be sold in ten or twenty years when it has been improved and has appreciated in value.
- Your will could transfer 50 per cent of the farm to your child's trust on your death. Your child could agree to forego any beneficial interest in the family trust after that point.
- The farming child may progressively buy out the siblings – after all, the siblings may have little interest in continuing to be beneficiaries who receive a little income from time to time when the farm makes a profit; i.e. they are beneficiaries of an asset that they don't own, cannot sell and get very little from.

You would appoint new trustees in your will. It would also be wise to write a letter to the trustees detailing your wishes; while this is not binding on the trustees it sets out in writing what you would like to have happen after your death.

EDUCATION

Parents always want to give all of their children a good start in life. Educational opportunities are needed by all of your kids no matter what they will choose to do in the future. Making sure that all of your children have the means to earn a good income and create a sure place in life is an obvious way to avoid some very difficult problems, such as having to take care of those who have not done well in the future. Some of your children may have the ability and drive to take their careers to an advanced level. If you are contributing to their further education or to setting them up in careers or businesses it should be made clear to all that this will be taken into account in the future. I can remember a client telling me his father called him into the study one day and said, 'My boy, your brother Jack will inherit the farm. You will get a first-class overseas education!' And that is what happened. It can seem patronising, non-consultative and even downright unfair by modern standards – and given the size and wealth of the property involved it probably was. (However, I also know that the brother who inherited made substantial contributions to the other siblings later.)

I have been present at a number of family conferences where adult children appeared to 'forget' all of the money that had been spent on them to date! Be clear with teenagers about what you will and will not fund as they approach their post-school education. My own opinion is that it is important that young people contribute in some way, such as by funding living costs with holiday work while you pay fees. As they progress there may be further choices and expenses, such as graduate degrees or qualifications, overseas experience, the option to buy shares in a business or the purchasing of partnerships. I would recommend that you deal with all of these opportunities in a frank and open way and commit the details to paper. No matter what your circumstances, each of your children should be aware of and recognise the investment that is made in them no matter how willingly this is given.

Life insurance

You can purchase whole-of-life insurance policies that will have a surrender value at some date in the future. The trick with these policies is not to cash them in early. This is a 'set and forget' strategy in order to have a specified amount of capital some time in the future. You can probably create more wealth by undertaking your own off-farm investment – but that takes time and effort. (More about that later – see chapter 8, 'Investing off-farm'.)

Life insurance offers the usual peace of mind. In addition you know that there will be money to pay off-farm debt when you go should it be needed. It can also provide inheritance money for other siblings. This may give you much more freedom in retirement as you can count on this money being available for legacy.

Subdivision

Sometimes a sibling really wants a piece of 'home' that they will be entitled to forever! Your farm may be able to subdivide a piece of land that allows one or more siblings to build a home or holiday house. I have seen this idea taken further with subdivided land being used for urban development purposes. You may want to use the farm to leverage such subdivisions and developments.

No magic bullet

Farms, like other family businesses, have a problem in that a great part of the family wealth is likely to be tied up in a single asset. No one solution exists in that all families are different and circumstances continue to change. The best approach is to start early and adopt a flexible strategy that can accommodate changes as they arise in your family and wider circumstances. And you can always resort to 'Russian roulette'!

Russian roulette

It goes without saying that there can be a great deal of emotion surrounding the succession of the farm and other special assets that the family may have such as a homestead, bach or crib, antiques and other family heirlooms. Many farmers are distraught at the prospect of their children rending the family fabric as they squabble over who gets what. Others anticipate that the real trouble may only begin when they are gone and are horrified at the thought of schism in the family. This of course only gets more complicated as children bring partners and grandchildren into the family group. One way of resolving some of this is to play Russian roulette. The children can 'bid' blindly for the asset that they want and the highest bid wins. It's hard to argue with and even harder to carry a grievance against: the one who wants it most and is willing to pay for it gets it, and the others presumably benefit now – or eventually – from the introduced funds.

These scenarios will hopefully give you some ideas to consider as you read on.

Chapter 4

Family meetings

As I have stressed, I believe that family meetings are very useful – even essential – for running a healthy family business. No matter what stage your succession process is at – whether you and your children are relatively young and expect succession to be decades away, or whether you are close to your last fencepost – it is probably worth starting to hold such meetings. They will be especially useful as the rest of the family begins to consider succession and the transitions that it may mean for the family and the farming business.

The value of a family meeting is that it brings everyone involved together to share their dreams and goals, discuss problems in the family and business and agree on plans for dealing with the issues that arise. Even if some family members are very young and not yet active in the farming business, these meetings can be a great way for them to learn about the farm and the family's traditions and for you to inculcate the values that you hold dear. Family meetings also help foster a sense of identity and respect for the customs that you may want to preserve.

I have been involved in many family business meetings over the years. I think they are very useful in helping build both stronger families and better businesses. Most importantly they give family members a chance to address the future in a structured way. By their very nature, formal meetings encourage people to think about what should be on the agenda, to consider their respective points of view and to have an orderly discussion about even the most divisive of issues. If family meetings are held regularly then things do not go unaddressed and culminate in an outburst of frustration and emotion – not usually desirable when the participants still have to live together as family! Formal meetings encourage people to anticipate issues; when you discuss development of an individual, for example, people are prompted to think about the development of everyone and whether or not the family is anticipating future needs.

Remember that a family meeting does not imply a democracy: the farming couple are still the owners and probably also the main operators and should be making the key decisions. However, there may be other people involved, such as brothers and sisters of the owners, inactive shareholders and growing or adult children. Every family is unique and different things may need to be discussed at the meetings; but at the very least the meeting can be informative and help mould the kind of family business culture that you want. The real value is seen when the farm has to deal with significant changes. These could be related to the business strategy or simply to the kinds of changes that will be occasioned by the farming couple's decision to do less or retire altogether. Regular meetings over the years mean that none of this comes as a shock and everyone's thinking is being moved along and kept up to date all of the time. It also reduces the risk of family members being unaware of what others in the family think or want.

WHAT SHOULD BE COVERED IN MEETINGS?

There are certain things that everyone will benefit from understanding.

Budgets

How does the family farm work? What really earns money around here? How do we allocate money? How does this affect all of our lives and choices? As you move towards a transition it is essential that everyone understands the levels of income and profit in the farm so that realistic decisions can be reached about how the farm will continue to operate and what money can realistically be made available for retirement options.

Capital expenditure

What money has been invested in the farm? What needs to be done to keep it healthy? What will it cost us to become more productive? As you progress towards retirement you may need to consider what capital can be taken from the farm by selling it outright or selling down to a successor.

Off-farm investments

What money has been invested off-farm? Why did we make these investments? How are they doing? How do we manage them? What else are we considering for investment? As you progress there are stronger and stronger arguments for investing off-farm in order to diversify and facilitate succession issues for both generations. This is a great area in which to get younger people involved in order to build up their competence and allow them to showcase their abilities.

Heritage
What has happened on this land or in this business? Who led the changes? Who has invested their lives in this business? What do we as a family hold dear and want to preserve? What are we able to change without losing what is precious to us? How do younger family members view what they have received by virtue of being born into our family? For what are they willing to fight to keep for the future?

Leadership and management
Who does what? Where does the vision and passion come from? Who does the hard thinking and shoulders the tough decisions? Family meetings are a great way to illustrate what it takes to succeed on a farm. They help people sort out the realities from the fantasies; they also showcase the real thinkers and workers. Meetings are a great way to help people see who has the abilities and skills to run the farm well and to be a responsible manager – perhaps on behalf of other interested parties.

Education
What are we good at? Where do we need external advice? Who brings us this help and information? What else do we need to know to be good at our business and to future-proof it? Over time, members of the family get used to the upskilling process that is needed to keep the business healthy. They also get to know the various advisers that assist the family. When approaching succession this helps keep the family open to outside help. Outside advisers add objectivity and help the family frame the right questions rather than just focusing inwards on family matters.

Policies
What has happened here and how have we dealt with it? What is likely to happen and what problems or decisions might we face in the future? The things you value and the experience and learning

you have accumulated usually mean you have views about how you should make decisions for the future. Farming families will have views about how money is spent, about investment, about education and about use of external advisers. It's a good idea to make these policies explicit rather than having everyone try to second guess which way the farming couple will jump. It's quite simple and beneficial, for example, to have a policy that says we will fund post-school education or off-farm training for all family members or all farm employees. Over time the family can discuss and develop policy around issues that are likely to arise or become contentious, such as the involvement of children's partners. Being clear about your views and discussing some of the 'what if' scenarios that can arise in family businesses can both educate everyone and potentially stave off conflict. Obviously it's best to keep a written record. You may gather useful items for discussion by observing what happens in other farming families and raising these issues under the banner 'What would we do if…?' Younger people learn great process skills in this way. It does not prevent difficult matters from arising but rather gives people the skills to address them and a vehicle for making better decisions.

One of the advantages of instituting family meetings is that succession issues come up naturally and in a timely way. From time to time family meetings will traverse such questions as:

- How did this farming business begin?
- How did the present owners get the farm?
- What has been done to develop the farm?
- What is the vision for the future of the farm?
- What goals have been set?
- What does the business plan say?
- Who are the key operators of the farm?
- How have we developed these people?
- How are we developing people for the future?

- Who is emerging as a leading thinker?
- Who are the leading managers and operators?
- What critical choices does this farming business face?
- When will we face them?
- What does each key player think?
- What does each family member want?
- What conflicts do we have?
- What options do we have?

It is highly likely that family members will have an insight into these questions (and answers) if the family has been holding meetings for a while.

How to run family meetings

It goes without saying that not everyone gets what they want in any business meeting – family business or not. A meeting does not imply consensus. Looking for consensus often results in wheel spinning or unfortunate compromises. It's obviously desirable – but often not possible – that everyone is happy with the decisions made. However, inclusion in some way over time means that everyone understands why and how decisions were reached. Participation is what is needed – each person has had a chance to have a say, if not a vote. This goes a long way towards building understanding and keeping people aligned, and the evidence shows that people generally support decisions in which they have participated. The key difference in a family business is that you are still family when all the business is done, and most families want to avoid rifts – even if everyone is not entirely satisfied with the decisions that are made. I think that there is a lot to be said for open management in any business – even in a large corporate. And I am absolutely convinced that it is the best way to go when a family is in business. At the very least you are giving your family a great education. At best you are laying a solid foundation for the future and building support for whatever comes next. You can't lose.

Getting the most from meetings

You can do a number of things to make family meetings as effective as possible:

Start early in family life
Initially the children's only interest may be in future plans for holidays or how Mum or Dad will organise to be free to participate in sport or school events! You might choose to include them only for ten minutes or so. As time goes on, they can remain for more of the meeting, be given more information and be asked for more input.

Have an agenda
The agenda should be circulated in advance. Important issues need to be signalled early; you do not, for example, spring a discussion on selling or succession on the participants at the end of a meeting!

Keep information appropriate
What do the participants need before and during the meeting? Too little and too much are common sins.

Get a big table!
Symbolism matters and it's much better to get everyone around the table rather than sprawled around the living room or being talked at classroom style. A great big table – round if possible – implies that everyone has a place and it doesn't feel so hierarchical.

Chair the meeting properly
It's not a given that the owner should do this. An external person such as the farm accountant or a facilitator can be very useful for providing some organisation and focus. Because they are not there to air an opinion, they can focus on keeping order, making sure that all contributions are heard and ensuring that the discussions don't

get off track. You might rotate the chair as younger people become more skilled: it builds self-respect, as well as respect for each other and the process, and they learn how hard it is to do it well!

Involve outsiders

There are several people who might attend all, some or occasional meetings. Consider when you should involve your trustees, farm adviser, accountant, lawyer, investment advisers and wider family members. The benefits work both ways.

Even if you have never held a family meeting before, you may want to consider doing so as you approach retirement or succession.

A special family meeting

The lifelong habit of family meetings as described in the last chapter is ideal. However, often it's too late for that: the family has had years of little or no communication or consultation about the future of the farm or the intentions and wishes of the individuals involved. People often avoid the difficult conversations over the years – perhaps because they are fearful of what they may hear, or simply because opening up some of these topics is just plain *difficult*. It's understandable that people worry about what to say and how to say it.

But the issues do not go away. Sooner or later, the farming couple have to make some decisions about the future. At some point you are unable actively to farm – the job is too big, the task too physically difficult – or you just want to do other things. A decision to sell or pass on the farm is inevitable. Now all of the conversations that should have taken place over the years need to happen with some urgency.

The farm belongs to the farming couple. They can make whatever decision they like about their own future and the future of the farm. However, most people in my experience want all of their family to be happy about – or at least to understand – the choices they make. It's important to get everyone together for these discussions. My observation is that it is best if the children's spouses and partners do not attend – this is about the primary family.

First steps

The first step is to call a meeting. I think it is helpful to get some external assistance for this. An external facilitator can help you draw up an agenda. You will have things you want to present and issues you want addressed. The facilitator can also canvass family members to ascertain some of the issues that need to be raised; being consulted like this helps people feel included and also prepares them for the meeting. Interviewing individuals in advance means that they feel respected and also takes some of the heat out of the meeting. Many people may be able to assist you with organising and managing the meeting. Your farm adviser, accountant or lawyer may be suitable – or may recommend someone who is good at this kind of work. You may have a close friend or family member who could do it well, or one of your trustees may be ideal. I find when I do this kind of work that clients appreciate the fact that I am completely independent and able to look at the issues with fresh eyes.

Everyone should know why they are attending the meeting and what will be discussed. Once again there should be an agenda and it should be circulated in advance. You do not want people to feel 'ambushed', which is how they will feel if they have not had time to consider the issues beforehand.

What to cover in the meeting

The meeting needs to cover several issues. Each family is different but if you are approaching succession some time soon you would expect to discuss many of the following questions:

- What do I want?
- What values do I hold dear?
- What would I like to see happen to the farm?
- How do we make sure Mum and Dad have enough for the rest of their lives?
- What are the arguments for selling up?
- Who should succeed?
- What will happen for everyone else?
- How will we deal with our siblings?
- What are the family trust arrangements?
- What's in the letter of wishes?
- What is the will going to say?
- What are the timeframes?
- What about our spouses and partners?
- What if the successor becomes divorced?
- Where to from here?

Speaking and hearing

It is also important that everyone speaks on each issue: there should be no opportunity for people to say afterwards that they were not consulted. Having everyone present means that each person hears the views of everyone else. This is much more effective than having several separate conversations that are relayed around the family – a recipe for confusion and misrepresentation. An outsider can be most useful in facilitating this exchange. A truly independent but skilled outsider can also ask 'naïve' or obvious questions – such as, 'What do *you* want?' Although of course these questions are obvious, the very reason that the meeting is being held is that they have not been asked or discussed previously. It is much easier for an outsider to ask these questions of each person in the room. An outsider can be presumed to be interested, and it is not assumed that they already know the answer – as it will be for many family members.

People tend to support those decisions to which they have contributed – even if they did not get their own way. And good communication and the sense of being consulted get rid of a great number of ills, such as bitterness and hostility. While some may not get the result that they wished for they nevertheless know that they had their say and their views were heard. There's nothing like a nice big asset such as a farm or business to divide a family as there's a lot of wealth at stake! Members of the younger generation may have a sense of entitlement and members of the older generation may bring a weight of expectations – reasonable and unreasonable – regarding the younger people.

Get good help

A facilitator can also make sure that all the necessary information and records are available for the meeting. Members of the family will know you are serious because you have involved an external facilitator. They are also more likely to behave appropriately if there is an outsider present; it is the fear of family ructions that stops many people progressing their succession plans.

I have been involved in many such family meetings over the years. There are several advantages in having someone external manage the meeting. The most important one is that the external person is independent, and brings no agenda other than the resolution of all the issues in such a way that everyone feels they have been heard. A truly independent person has no vested interest in either the farm or the family, but rather has an interest in the welfare of both. Everyone else in the room is involved or committed in some way and will see and hear each contribution and proposal from a personal point of view. An external facilitator also has the job of keeping order, and can do this by having an agreed agenda that allows everybody a turn but does not allow one party to dominate. It is so often the case that families behave better when there are outsiders present. When emotions run high or debate becomes heated a good facilitator can model problem and conflict resolution and assist family members to have appropriate conversations. This can defuse potential rifts – rifts which no one wants. A skilled facilitator has the advantage of having seen it all before, but also of being truly independent – with no 'skin' in the game.

It may be appropriate to have your lawyer, accountant and trustees present as well. They have professional knowledge regarding the issues and are also familiar with the specific history of the farm and family. My experience is that it is better to have them as specialists in the room rather than as facilitators; it's all too easy for them to be seen as defending the actions – or inaction – of the past. It's often best if the family members meet the first time and the specialist advisers join them for subsequent meetings, as needed.

Remember everything does not have to be decided in one meeting. In my experience what is most important is that the meeting is actually held, the issues are broached and each person speaks and is listened to with respect. Many questions may arise and there may be some debate about the facts. It is useful to have to hand the documentation that you may need, e.g. records of other money that has been distributed, valuations and copies of trust deeds, but it may take some time to follow up on some of the issues. It is worth taking the time to do things properly – especially if your principal goal is family understanding and ongoing harmony.

The facilitator can be charged with managing follow-up and more meetings may be needed. It is my experience that family members value the opportunity to get the issues on the table, air their views and hear what others have to say 'straight from the horse's mouth'. Many families comment that they wish they had done this several years previously and avoided the stress, undercurrents and second-guessing that has marred family life.

Often, the single biggest issue that must be resolved by the farming couple is appointment of a successor.

CHAPTER 5

Identifying and preparing your successor

IDENTIFYING SUCCESSORS

It may not be clear at the outset who will end up running the farm. There may be no potential successor in which case you will concentrate on the finances and plan to sell when you are ready. You may be lucky enough to have more than one child who is interested in farming. It is important to approach succession with an open mind, especially when your children are still relatively young. To state the obvious, parents are often poor judges of their children's aptitudes and competence. Parents' impressions sit at both ends of the spectrum – they are either blind to their offsprings' abilities or, alternatively, believe that they walk on water! The rest of the world is not so blind.

There are several possible scenarios: one child may be passionate about the farm but have less aptitude than another; more than one child may want the farm which is not large enough to sustain two or more families; a daughter may be the best candidate but your sons (and everyone else) expect a male to succeed. The farming couple may not even agree on the best successor. Choosing between children can be very difficult as someone may get hurt no matter what you do and that goes against all the instincts of a parent. If you begin early enough it is relatively easy to solve these problems by growing the size of your farm, for instance, so that siblings can farm in partnership; investing in another farm for a sibling; or securing an off-farm investment that better suits another sibling's talents.

Procrastination costs

Procrastinating about such difficulties often means poor outcomes. The situation may reach paralysis for a matter of years as you respond to the dilemmas by doing nothing – and even worse, not talking about it either. The succession problems do not go away – but unfortunately the 'kids' often do! I know of a highly competent sixty-year-old who is still waiting for his ninety-year-old father to make a decision and hand over the reins! However, I have also seen many young people give up, move away and get on with other careers. It is highly unlikely in the twenty-first century that young people are going to put their lives on hold while this kind of dithering – even paralysis – is allowed to constrain everyone's lives.

People who can't bear the indecision often make another poor choice: they rush into nominating a successor and do so on emotional or sentimental grounds. So the 'eldest son' or the 'one who needs it most' is appointed rather than the successor who is best suited and prepared for the role.

I think it is wise to take your time and employ a structured approach. Whether there is one candidate or more, the period of preparation for succession should be an extended one.

Make no assumptions

I think it best to begin by deciding to make no assumptions. The family can be a strange institution. (I write with some authority!) One of the most fascinating aspects of family life is the way family members make assumptions about each other over the years – and almost never check these assumptions out.

This can be particularly poignant in the area of succession. I have dealt with several farming families where parents assumed that a particular son would take over the farm only to find out (far too late) that he had no wish to do so. In another case, I knew of a daughter who had a passionate desire to farm but was never even canvassed regarding her wishes for the future of the farm, much less considered as a possible successor. These sorts of unchallenged assumptions can have devastating consequences, especially if they have been let run for years. It is very important to know what each family member thinks and feels even if you do not agree or will find it difficult to fulfil all their dreams. As for any other business, good communication is essential to success in farming and I have already stressed the importance of starting this conversation as early as possible with each of your children. Talking with your children about their interests and career aspirations is important in any family but it is clearly critical within a family business. These conversations need to be maintained over the years – we all know how young people's interests and capabilities change over time.

What is the model?

It is also important to be conscious of the model you and your spouse provide. I have talked with a number of farmers who were distraught that none of their children seemed interested in the farm. I asked these farmers, 'What are you modelling?' This often led to knowing looks from the spouse; after all, if the children have observed constant complaining about farming conditions, relentless hard work and a joyless existence on the part of their parents, it is not surprising that they can't wait to get away to pursue another career! To their great credit, this conversation has been the catalyst for several farmers to look at how they are approaching the business, how they are organising their family life and whether, for example, they are having regular family holidays. Passing on the farm to family is only worthwhile if you have found it to be a fulfilling life and the next generation is likely to experience the same satisfaction.

In many families it will be obvious who is going to take over the farm. There may be only one child and it may have always been openly agreed that this child will succeed. Even in families with several children the successor may have self-selected early on and everyone else is in agreement. Sometimes it's not so obvious – and not so easy. You may have more than one child who wants to succeed and appears suitable. Alternatively you may have more than one child who would like to farm but you may be undecided about suitability. Again, each family is different and there will be different issues to address.

What does it take?

It really can help to take the most objective approach that you can. Even though it's all about your children and your farm it's actually much easier to think clearly if you start with the role itself. What is required in terms of knowledge, attitude, skills and habitual behaviours (KASH) to succeed in the role? You and your spouse will already be quite expert in identifying what it takes to run your farm well and have a satisfying life in your community; this is a picture or role description that you will build over time. Other people can help with this – you may have good friends or neighbours who are experienced in farming similar farms and who might have valuable contributions to make. Your advisers can also have an input – they will have a different perspective on what it takes to succeed and they have the advantage of dealing with many farmers. You will also want to know their views on what it will take to succeed in coming years – years which will present different challenges from those faced during your tenure.

TABLE 1: KASH – Knowledge, Attitudes, Skills and Habits

Knowledge
What do you need to know in order to run the farm well in the future?
Attitudes
What attitudes are required to be a successful farmer?
Skills
What do you need to be able to do well in order to succeed in farming life?
Habits
What kinds of habits do you need to acquire to develop the necessary character?

Defining the elements required to succeed in the role automatically leads you to assess each of your possible successors and put development plans in place.

Preparing successors

It is difficult to be objective about your own beloved children. However, if you have a framework as discussed it is much easier to identify strengths and weaknesses and take a structured approach to preparing successors. I believe that there should be an individualised plan for each potential successor, remembering that things will be very different in thirty years. Do remember too that almost every master was once a disaster!

There are five key steps for developing these plans and making them work:

- establish what is needed for success;
- assess each potential successor against the needs of the role;
- devise a programme for each person;
- implement the programme;
- monitor progress.

Step 1: *Establish what is needed for success*

This is a very useful exercise for the farming business itself. What does it take to run this farm well? You will be aware of the operational requirements and so will have a good sense of the essential knowledge base and skill levels. You may also have a feel for what is desirable – maybe including knowledge and skills that you do not have. It is worthwhile to involve others, such as your partner, accountant or farm adviser, or a respected neighbouring farmer. The reason for this is that in all likelihood you will be unaware of some of the knowledge and skills you have that you take for granted. Others will not be blind to what you bring to the job. This can be especially so when it comes to describing the necessary attitudes and habits: you may take your vision, application, hard-working habits and resilience for granted but others will not. Often these attributes are more important for success than the obvious knowledge and skills component; frequently we are far less aware of our own character and how it contributes to our success than are others. So it is wise to enlist the help of those who well understand what is needed and who will have a view on what has or has not worked effectively on the farm to date.

Be sure to look ahead as well: it may be that what is needed for future success differs somewhat from what was required during your tenure. What's on the horizon for farming in your region or in your specialist area? What market conditions can you expect in the coming years? What will be the big challenges facing your successor? For what do you wish you had been better prepared? Remember that you must prepare your successor for *tomorrow* rather than yesterday – just think about how different things were thirty years ago. You will have seen a lot of changes during your tenure; it is likely that your successor will face even more.

TABLE 2: KASH needed for success

Knowledge
What do you need to know in order to run the farm well in the future?
General knowledge, e.g.: • how business works; • the basics of human resources management; • finance; • trends in farming, marketing, technology. Specific knowledge, e.g.: • how this farm works.

Attitudes
What attitudes are required to be a successful farmer?

You must be, e.g.:
- goal oriented;
- accountable;
- responsible;
- enterprising;
- a competent steward;
- clear in your values.

Specifically, you must be, e.g.:
- skilled in collaboration.

Skills
What do you need to be able to do well in order to succeed in farming life?

You must be good at, e.g.:
- speaking;
- writing;
- numeracy;
- finances;
- negotiating;
- organising.

Specifically, you must be, e.g.:
- good at pasture management.

Habits
What kinds of habits do you need to acquire to develop the necessary character?

You will need to demonstrate excellent, e.g.:
- work habits;
- personal discipline;
- habits of mind;
- commitment.

Specifically, you will need to demonstrate, e.g.:
- good judgment.

Step 2: *Assess each potential successor against the needs of the role*
The task here is to assess the development to date of your potential successors. You are trying to get a fix on what they know, and what they are able to do. You are also trying to become objectively aware of their attitude to life and work and to observe what their personal management habits are like.

Again, outsiders can offer some help. You may have teachers in mind who have taken a close interest in your son or daughter and can see the emerging competencies and character clearly but through a different set of eyes. As before, the more your farm advisers interact with your family and attend family meetings the better they will be able to form a view and give you good and useful feedback. Obviously the earlier you begin to observe your children in this way the easier it is to be objective and identify where there is work to do to prepare your potential successor.

This exercise gives you a baseline. It also provides you with something against which to measure progress. Best of all it lends some objectivity to the exercise – if you and those helping you can agree on the current level of readiness of each candidate then you are far less likely to fall prey to emotional decision-making as time goes on. These exercises often throw up unwelcome information. It may be that your 'preferred' candidate at this stage doesn't look very promising after you do this assessment. However, 'the facts are friendly'. At least you can now confront the problem. The more time you have in hand the less the knowledge and skill deficits matter. Poor habits take time to change too and a bad attitude, such as an assumption about succession or an unwillingness to work hard at anything, will need to be confronted immediately. There is usually a mixture of good and bad news but at least you have data and an objective process involving several others. This makes it far easier to take action as required. Ignoring problems now only leads to crises at a later stage. Incidentally, everyone else almost certainly knew about these 'problems' before now. See this as a chance to flush out information and use it well for decision-making.

TABLE 3: Assessing the KASH of your successor

KASH needed	Examples of KASH successor may have or may need to acquire
Knowledge What do you need to know in order to run the farm well in the future?	*Understands the farm well;* needs to develop an understanding of farming business principles and gain wider experience in dairy farming
Attitudes What attitudes are required to be a successful farmer?	*Enthusiastic and enterprising;* needs to learn to take a more collaborative approach and better appreciate how the family and farm intersect
Skills What do you need to be able to do well in order to succeed in farming life?	*Demonstrates good stock handling and pasture management techniques;* needs to develop planning, decision-making and financial skills
Habits What kinds of habits do you need to acquire to develop the necessary character?	*Diligent and committed to the farm and family;* needs to develop wider interests and involvement beyond the farm for balance and perspective

STEP 3: *Devise a programme for each person*
The next step is to ask what needs to be done to develop the necessary KASH in each of your potential successors. Obviously, the more time you have in hand the less it matters how much development is needed. If your preferred successor is young and you have a decade or more then there is plenty of time for the person to acquire the knowledge and skills needed. Obviously if the person is older you have to ask why they are still lacking the KASH they need to succeed and whether or not there is time to remedy the situation. Poor attitudes such as lack of interest or ambition are clearly much more difficult – if not impossible – to address. And what of habits and behaviour? Sometimes

bad habits are just that – people often learn to do the wrong thing or do things poorly. This can show up in everything from making soft decisions to having poor personal management skills around routines and tasks. But just as bad habits are formed over time so can better new habits be learned. It's not easy to change behaviour but if the right attitude exists and if there is sufficient motivation and direction then all the right ingredients are there.

Identify what knowledge and which skills they need to develop. Ask which attitudes and habits need to be nurtured. Each candidate will have a different set of needs. The key question here is how best to use the time you have left to prepare the potential successor(s). What will they do to learn the knowledge and skills they need? How will you build the appropriate attitudes and habits? Write down all the suggestions from your group and work out a schedule for each. Some of this will be practical, on-the-job learning while some will be theoretical and will require classroom training, reading or specific assignments.

TABLE 4: Sample KASH development plan

KASH needed	Suggestions for development
Knowledge What do you need to know in order to run the farm well in the future?	Your successor could: • take a course on an accounting package such as MYOB; • visit similar farms in the region; • take an overseas trip including farm work, supermarket visits, etc.
Attitudes What attitudes are required to be a successful farmer?	Your successor could: • become involved in local not-for-profit activities; • consult a mentor with particular emphasis on growing independence and challenging thinking.

Skills What do you need to be able to do well in order to succeed in farming life?	Your successor could: • attend a course to develop meeting skills; • practise meeting skills by chairing family meetings; • improve planning and analysis skills through specific delegated farm projects.
Habits What kinds of habits do you need to acquire to develop the necessary character?	You could: • delegate full profit and loss accountability for a section of the farm.

STEP 4: *Implement the programme*

You should be quite explicit with each person you are developing. Explain that you are putting a plan in place to grow skills and competence. It may be wise to emphasise that this does not mean they are being promised the farm: it's never a good idea to create Crown Princes or Princesses! What you can promise is your full commitment to the development programme; this learning will serve the candidates well whatever happens ultimately. While not setting up a competition, you should make it clear that you will be observing their progress closely and they will have to demonstrate their capability to you and all of the others concerned. In my experience young people in all situations crave this support and direction: far from resenting it, they relish it and expect to be challenged. Members of this generation know how competitive life is; they know that if they are not developing and moving forward all the time then they are, in fact, going backwards. No matter what they end up doing with their lives this process will serve them well.

Step 5: *Monitor progress*
It is important to follow through on all of this work. The prospects need to know that you haven't put all this in place in a great burst of enthusiasm and then forgotten about it. I recommend regular reviews where you (and your group) look at what has been achieved and then update the plans. Obviously the plans only need to be detailed for the coming period – things that need to be done in a few years time can wait for further clarification later.

Be frank with your children about their progress. This needs to be done on an ongoing basis – once a year is far too infrequent. The younger the candidate, the more frequently you will need to observe, give specific feedback and review what is happening. The young people should get continual feedback from you and other coaches, teachers and mentors. Continuing to practise the wrong stuff is the worst of all processes. And do remember that learning by doing has its limitations. First of all, it's a 'sink or swim' method and that's very wasteful and can be destructive for the people you burn off. Secondly, learning by doing takes far too long: why would you want these young people to make all the mistakes in the world in order to learn what someone could easily have taught them? All the modes of learning – formal classroom tuition, reading, observation and experiential learning – have their place.

Developing people is never wasted. You have nothing to lose and everything to gain by taking an objective and structured approach to developing your children. If you have several possible successors then they are all worth the time and effort. All will benefit from the planning, nurturing and learning. Knowledge and skills never go to waste. I don't believe in setting people up in adversarial competition with each other: it only creates winners and losers and, given that all involved will still be part of your family, or very close to it, it would not make good sense. On the other hand it does not do young people any harm to understand that their succession is not a foregone conclusion and that you have other options. The striving and testing that is implicit in a development programme is a very good way for a prospective successor to show what they are made of, meet the challenges that it brings and learn a little humility along the way. Younger people can succeed to management roles long before any ownership is transferred – you may well delegate significant responsibilities quite early. One of the great benefits of a well-structured development programme is that everyone gets to see the person work hard to meet the targets and expectations: there will never be any doubt that your successor has done the hard yards and is a worthy candidate for the job. This is especially important if staff or siblings will have to work alongside this person in the future – they need to be sure that your successor's worth has been proved and the role is deserved.

Preparing your successor is one of the most important jobs you will have as you move towards putting in your last fencepost. The earlier you begin, the easier it will be for everyone and the greater your chances of a pleasing outcome. But it's never too late and there are many others who can help. And it's a really good idea to have external people involved – they bring expertise in other areas. They are also more objective: to a certain extent your child is always a nine-year-old in your eyes! Your key task is to start (or continue) the process no matter what stage of succession you are at when you read this book.

There are also several other crucial aspects of change to explore.

Chapter 6

Getting ready for change

It is hard to make a change – they say that the only one who likes it is a wet baby! Breaking the habits and routines of a lifetime is never going to be easy. Apart from cataclysmic change forced upon us by unexpected deaths and other disasters most change starts in our own minds.

One of the positive aspects of farming succession is that we all know that it has to come – some time. Depending on our personal inclinations we have decades to consider what we want and time to get everything in place to ensure that we achieve most of our wishes. Hopefully you have been thinking and talking about aspects of this for a long time and have had a chance to say, 'One day I'd really like to…' or 'When we stop full-time farming I am going to…' Nevertheless, many people find it hard to make the break no matter how much they want to do other things with their time.

You are clearly the most important actors in this drama. Hopefully you have been talking all of your farming lives about what you want, how succession will be organised and how your financial affairs will be prepared for these changes. Not all couples are good communicators, however. We all learn these patterns of communication in our own childhoods and we can inherit a tendency to avoid discussing anything that is likely to expose differences of opinion or lead to conflict of any kind. There's a very real possibility that this is the case for at least one of the partners: many of us were brought up to get on with things rather than talk about them and to put the needs of others (the farm, the children) before our own. This is not a good preparation for the discussions that need to take place now.

The best strategy is to start as early as you can – and that's today if you haven't already begun. One partner almost always initiates these discussions. The talk may be triggered by events in the family, changes on the farm or things that happen to a relative or someone in the community. You may find this a little stressful – and we all have a tendency to avoid the difficult conversations! This may be especially so if you feel that you and your partner are misaligned as far as succession issues go: she wants to retire to Brisbane, he wants to move 500 metres down the road; he wishes to stop working, she dreads the prospect of his hanging around her house; she wants to visit the art galleries of Italy for their world trip, he wants to follow the All Blacks on tour! But sooner or later you must confront the issues and start talking – or life makes all your decisions for you. Not deciding is a decision – usually with unfortunate consequences.

What does each of you want?

Start to think and talk about what each of you wants for future years. What are your dreams? What do you want to have, to do, to be? Where do you want to live? How would you like to spend your time? What has been on hold all of these years as you struggled to develop a property and bring up a family? What did you dream of when you were young and do these dreams still have potency? What are you still wanting to achieve? What is the legacy (a much bigger idea than just money!) you wish to leave? There are many options to consider here because there will be sufficient money in a farm to give you lots of choices – unlike in many families where there are few choices or no choices at all. You have the opportunity to design a whole new life – your own succession to something new. And it's much easier to move towards something you want rather than worry about moving away from something familiar, whether satisfactory or not.

I think it works best if each partner considers these questions separately and makes some notes. Writing it down means you are serious and it also prevents a more vocal or articulate partner from ignoring the other's dreams.

TABLE 5: Dreaming exercises

1	Imagine that you are living the ideal life:
	• Where would you be?
	• What would you be doing?
	• Who would you be with?
	• How would you spend your time?
	• What would a typical day be like?
	Write about it; draw it with big crayons; build a collage; set it to music; write a poem – whatever medium appeals to you.

2	If you won Lotto…
	After you'd spent a month on the beach in the Bahamas what would you do next?
	Write all your ideas down – no matter how outrageous!

3	If you had a year to live how would you spend it?

4 If you knew you'd live for another hundred years what would you do?

5 Imagine that you are eighty years old...

Write a letter to yourself today from that future eighty-year-old. Tell yourself what you need to do; what you'll regret if you don't do it; how powerful you are; and how easily you'll accomplish your dreams...

6 What did you dream about years ago?

Why did you give the dreams away? What has been shelved? What dreams would you like to 'dust off' and breathe life into again? What feels like it would really be 'you'?

7 Write a list of 'Ten things I want to do before I die'.

8 Write a list of ten 'round-to-its'.

We are all accustomed to talking about things we want to do, to have, to be. We often dismiss these things as impossible as we grow up and become 'realistic' – or we put them on the back burner and talk of getting 'round to it' some day... What are your 'round-to-its'? Consider why you haven't done them so far.

9 Eightieth birthday

Imagine that it's your eightieth birthday. Your family has organised a surprise celebration for you and has invited your closest friends and colleagues from over the years to celebrate with you. Needless to say there are many toasts and impromptu speeches as the party progresses.

- What do they say about you?
- What do they remember that you demonstrated passion for?
- What personal qualities do they admire in you?
- What contributions do they say you have made?
- What difference do they say you have made to their lives?
- What particularly pleases you about what they say?
- What do you wish they could say?

10 You have twenty 'good' summers left...

What will you do with them?

Next you need to share some of what you have written; each person needs (respectfully) to listen. I find with clients that this is often the first time that their dreams have been aired. Most couples have been too busy over the last twenty or thirty years to spend any time talking about these dreams – and of course they can now come as a great surprise. Our values really matter – the goals that we care about so much that we are motivated to achieve and protect them. No matter what the marketers try to tell – or sell – us we are only truly happy when our deepest values are realised. So this is a time to share values and listen carefully. There's no right or wrong here – and no perfect answer. The best strategy is to make sure that each of you can have some of what you want – and hopefully there is considerable overlap.

Some conflict can be useful as it engenders discussion and the consideration of other options – just don't let minor differences take up all of your energy. If you no longer have anything in common then you will need to have a different conversation altogether!

Setting goals

Next you need to set some goals. I think the best way to approach this is to think about timeframes that are acceptable to both of you. When would you like to be able to ease back on full-time farming? When will you be prepared to sell or hand over responsibility to a successor? At what point if any do you want to be free of the farm entirely? While you may not be able to get exactly what you want at least you will have a feel for the number of years involved.

You also need to talk about what you will want to do then. Will you still be involved in the farm and/or do some work off-farm? Will you retire completely? Answering these and other questions about how you intend to live (see chapter 7, 'The numbers game') will allow you to 'price' your lifestyle in the future. So your main goals will be around how much capital/income you will need by a certain date.

Goals are essential to success. Most businesspeople, farmers included, have no difficulty in setting goals for the *business*. However, many of us are not so good at applying the same tried-and-true strategy to our personal dreams and values. One of my favourite definitions of goals is that they are 'wishes with due dates'! In other words, turning a vague dream or value into a goal requires you to be specific and forces you to set a timeframe. It is a great technique for making you keep your promises to yourself. Goals should be neither too easy to achieve, nor impossible. In fact, just like the children's story *Goldilocks and the Three Bears* goals should be 'just right' – not too hard nor too soft! Again, these goals should be in writing as we take what we write much more seriously. They will also need to be referred to and even revised often. Writing them down is a commitment in itself – especially if it is in your own handwriting.

Choosing strategies

Next you need to determine your strategies. What will you need to put in place so that you can begin to achieve these goals? Do you need to accelerate the preparation of a successor (as described)? Or does the farm need to be prepared for sale? Are there new skills either of you needs to acquire? If you are moving away, do you need to start looking at homes or land in another area in order to secure a hedge in property values? Are there interests you need to develop with a view to pursuing them later on? Many people leave it very late to get involved in sports, hobbies, not-for-profit groups or other areas that they would like to pursue. It's a good idea to start early to develop the areas of interest on which you expect to spend more time in the future.

The 'soft' stuff is paradoxically the hard stuff here. It is relatively easy to take any of the necessary actions one by one and you will be accustomed to writing business objectives and executing plans for your farming business. It is often much harder to turn the focus on yourselves and what you want and to begin discussing things that may have been put to one side for years.

You might find five key ideas useful to think about and kick around.

1 *This is not your parents' retirement!*
The whole idea of retirement and old-age pensions is credited to Bismarck in Germany in the late nineteenth century. It wasn't a big imposition on the state as the pension only kicked in at age seventy and life expectancy at the time was much lower than that. Why is this of interest? Well, many of us still view retirement as being relatively short and occurring close to the end of life. This is no longer true. But the long retirement is a new phenomenon and many of us do not have role models who have lived long, productive and fulfilling second halves. Many of us saw parents retire who were, compared to us at the same age, quite exhausted from a life of struggle that may have included at least one war.

But it's not like that any more. Chances are that you are considering easing back or retiring from full responsibility for the farm while you are still in the prime of life and in good health. You are almost certainly planning for a full second half rather than the end. So there are several mental readjustments to be made.

2 *It is not just a holiday*
Farming couples spend a lifetime struggling to get away from the farm for more than a week or two. It's just the nature of many farms – you may have no paid staff, your stock may need daily attention or there may be no one appropriate to take charge in your absence. However, the next period of your life – much anticipated as it may be – is not just a holiday. First of all it is likely to be long. No matter how much we long for lie-ins, long weekends or freedom from urgent tasks, the reality is that we will tire quickly of having

nothing meaningful to do. Holidays, after all, are enjoyable because they provide such a contrast. We use holidays to break the 'rules' of our working weeks. However, you are highly unlikely to want to do nothing for a decade (or two or three!) Nor are you likely to want to play golf, garden, see friends or just do very little day after day. It can be wise to plan for a month or two of this kind of relaxation when you first exit full-time farming but you really need to give some thought to what you would find engaging and meaningful over the coming years. If you still have some involvement with your farm (or another business or your investments) this may well fill the bill. But if you don't and you have no meaningful and satisfying commitments beyond your farming business then you should start as soon as possible to develop some. In my experience, talented, hard-working, busy people rarely enjoy being without meaningful and fulfilling 'work' (paid or unpaid) for long.

3 *Widen the circle*

Farming folk are often relatively isolated. Even if you have been quite involved in the local farming community these attachments may weaken once you are no longer farming full time. These issues are exacerbated if you choose to move away – even if it is to a holiday spot that you already visit frequently. We often don't notice that we make most of our friends and acquaintances through work, school (and our children) and within the family. When we are busy we often overlook the fact that we are no longer adding to this circle of friendships and interests. It gets harder and harder as we get older to meet new friends and develop sustaining relationships. I think it is important to recognise this well before you are likely to need a wider circle and to begin to build your networks – perhaps by volunteering, joining new groups and organisations or taking up new hobbies – well before you make a significant change in your life. If you are planning to move to another location, invest in building some networks and friendships there to ease the transition. As is typical in any relationship, one of you will be more extroverted and active beyond the farm than the other; pay particular attention to the person who is less likely to take the initiative to manage these changes actively. Again, it's good idea to write down some of the things to which you both aspire.

4 *Reinvigorate the relationship*

Over the years I have worked with many senior executives – mostly male. I have noted that many of the partners dreaded his impending retirement! They worried about him 'hanging around the house', 'needing to have lunch every day' and 'questioning where I am going and what I am doing'. It's quite funny at one level but taken more seriously it underlines what a change in routine is involved. It can also point to the fact that both partners have been busy and settled in their respective routines and these are about to be disrupted. Changes like retirement can put quite a strain on relationships even when they have previously been happy. Farming couples already see much more of each other than most couples but there will still be big changes afoot as you may well spend much more leisure time together. One partner may be quite reliant on the other initially for social contact and recreation. It's better to think and talk about some of this in advance rather than be shocked and upset about the reality at some future date. Over time, we all become aware of things about each other that really irritate. Now is also a good time to check out what adjustments each will need to make so that harmony for the future can be assured!

It would be wise to talk about some of the things you would like to do together and make some joint as well as separate plans for future times. Incidentally, this is a great time to take an overseas trip. Many couples defer travel until after they have moved on to the next stage. However, I think it's a great idea to go for an extended trip abroad as you plan for the new stage. It gives you time away which increases clarity and objectivity. It also allows for lots of talking time without the daily pressures of the farm and family. An added benefit is that it can be a real eye opener in terms of considering investments beyond the farm – after all, 99 per cent of the world's investment opportunities are outside New Zealand (more about that in chapter 8, 'Investing off-farm'). Go and visit people you know in other countries and talk about investment opportunities: far be it from me to suggest that this travel may be tax deductible…

5 *Wise rather than washed up!*
Each of you will have a wealth of knowledge and experience from your years of farming and your life's achievements. Many people are upset by the feelings of uselessness that often accompany doing less or having less responsibility. This can be personally debilitating. It can also translate into interference with the work of your successor, particularly if that person is a son or daughter, and that is not desirable either. If you have arranged to have some ongoing involvement in the farm *do* make sure that the ground rules are clear: it is most unfortunate if the generations end up resentful – or worse – because the boundaries have never been properly established.

There will also be many other avenues to explore as outlets for all of this knowledge and skill. Many community groups, schools and not-for-profit organisations would welcome the contribution that successful businesspeople and parents could bring to them. This can also be a great time to consider capturing much of your experience in other ways, such as in a history of the farm, family tree, reflection on the changes in farming over a lifetime or storybook for the grandchildren about the way things were. And don't delay! As with many other aspects of retirement, procrastination in laying the foundations for your future lifestyle can become very costly. Time is always your biggest ally when doing the groundwork in order to enjoy your plans.

You may well now be wondering how you are going to fund this wonderful future.

CHAPTER 7

The numbers game

How much is enough?

As you contemplate moving towards working less – or not at all – one of the larger issues to consider will be money. What will you have for income if you work less or stop working altogether? How much will you need? Where will it come from? This can be a particularly difficult question if you wish the farm to remain in the family – all kinds of issues may arise such as wanting to leave equity in the farm and having sufficient funds to treat other siblings fairly.

My experience is that people often overestimate how much wealth they need in order to scale back on work and live well on their investments. This is understandable and there are several reasons why you might think that you will never have enough:

You don't ever want to have enough!
This is really another way of saying that you are reluctant to make the change and move away from full-time farming. It is easy to convince ourselves that we need to accumulate capital for another few years or to acquire another million dollars as an excuse to put off any change.

You are unclear about the future you want
Generally, I find that the clients who know what they want are very easy to work with and it's also relatively easy to help them get what they want with what they have got. It's the ones who don't know what kind of future they would like who are stuck – and they often tend to use the 'money excuse' to avoid exploring what they truly desire.

You believe wealth should be grown but never spent!
Put like that it sounds ridiculous – and it is! However, you'd be surprised how common this belief is. We spend so much of our lives trying to accumulate wealth and grow our net worth that it is easy to forget what it is all for. It can be quite difficult to get your head around the idea of no longer growing wealth but actually using it for income and even consuming some of the capital. The habits of a lifetime can be hard to break.

You are an anxious person
This is normal and logical enough. After all, you will have spent a lifetime being anxious about weather, stock, markets and farming conditions. Anxiety is normal and we all need a certain amount just to survive. However many people are unduly frightened of running out of money before they die and so often argue for building up far more capital than they need. You may have been conditioned to believe that you need much more. After all, there are entire industries out there that have a vested interest in having you believe that you need more and that destitution is around the next corner!

You are scared of the unknown
You may have avoided dealing with any financial matters beyond the farm's business and may be reluctant to deal with the financial issues around your succession. Certainly, some capital will need to be reallocated and much of your wealth (from which you will get your future income) should be moved out of farming. However, all of this is relatively straightforward once you have decided what you want and, by extension, how much you need.

So how much do you need?

It is really important to put plenty of time into working out what you will need as you may want to leave some money in the farm or you may want to invest in another farm for a sibling or to give help now (rather than later) to yet another of your children. There is a lot of joy to be had from helping your children now when it is of most benefit to them.

Perhaps you do not need as much as you may think at first. After all, you may never intend to fully 'retire'. Most baby boomers are in very good health and have no desire to be put out to pasture for several decades before they die. While you may want to pass on all of the responsibility for the farm (and for early milking or mid-winter lambing!) you may well expect to play an active role on the farm for many years to come. In addition you are very well placed to be actively involved in farm management – much of which is done from the office with a computer nowadays. The time when you had to be young, fit and very strong to make a contribution on a farm is long gone.

Depending on what else you have created over the years you may also have off-farm investments that you would like to manage actively for years to come. You may also have skills that are in demand elsewhere – on other farms, in farming equity partnerships, in other businesses or in your community. All of these things make a big difference to the money you need as you are likely to have some continuing income.

Secondly, you may want to reconsider your expectations about legacies. Many parents no longer feel compelled to leave so much to their children. You may, after all, expect to live well into your eighties or beyond as life expectancies have risen a lot and are particularly high for those already reaching sixty. On that basis your children will be well established by the time you die and are unlikely to be in need of very much. Many members of this generation have been well taken care of as young people with advanced education and other opportunities – arguably they have been heavily invested in already and should have fewer expectations later. It is also unlikely that well-established, middle-aged children 'need' a large legacy. I do not believe that older people who have worked hard all of their lives should 'put off' easing their workload or constrain themselves in the second half of their lives because they feel that they 'should' be leaving substantial inheritances.

CALCULATING THE PRACTICALITIES

It is difficult to work out what you will need for the coming years. I tend towards the optimistic while many planners assume very low returns on all investments, higher levels of inflation and low- or no-growth portfolios, and calculate that you will leave your capital intact. This sort of thinking always points to the need for huge capital sums for retirement. However, many people looking to scale down their workload or retire altogether will live for another thirty or forty years. It is difficult to be predictive regarding such a long time period.

Some of the things you need to consider and which you can adjust to your particular circumstances include:

- *A home.* Many farming couples will have to buy a home for retirement if they are leaving the farm. How much will this cost? You can always choose to buy 'less' home, or be prepared to release equity in the house over time. Some people may be content to rent. Suffice to say that the cost of the home you choose to have has a big impact on how much is enough.

- *Lifestyle.* How will your living costs change and what do you expect to spend each year? It's important to do an indicative budget. Remember to recognise all of the costs that the farm may have covered, such as much of your food and fuel. Include a budget for leisure activities such as travel or recreation – you may not have had the time for such things in the past.
- *Earned income.* Is either of you expecting to continue to work? What do you expect to earn? Even relatively small amounts of earned income make a big difference to the capital sums you need. Obviously the younger you are, the more impact this has: if you are in your forties, and prepared to do some part-time or contract work over the next twenty or thirty years, then your capital needs now will be much lower than someone who wishes to stop working altogether and who can anticipate living for another forty or fifty years! And it's not just the amount earned: as long as you continue to do some work you tend to retain employability, thereby giving yourself a hedge. In other words, you can turn the tap back on should you ever need to. In addition, earned income tends to be relatively inflation-proof – if inflation soars, so too will the pay you receive which provides yet another hedge if you have a long timeframe to consider.
- *Other income.* At some point you will be eligible for New Zealand superannuation. Have you joined KiwiSaver? What income will you have from other investments?
- *Life expectancy.* How long will your money need to last? There's a big difference between planning for twenty and for forty-five years. You need to factor in how early you intend to

cease full-time farming, longevity in your family, your health status, etc.

- *Investment returns.* What level of return are you expecting before tax? As you age you will probably lower your risk profile (move more into bonds and cash) and therefore lower your expected returns. However, if you scale back while still relatively young you would still expect to have some growth in your portfolio.
- *Legacies.* The less you are committed to leaving, the more capital you can spend. Similarly, the more you wish to leave in the farm or give to other siblings the more you will need to have in order to sustain yourselves as well.

All of these issues should be discussed in detail with your partner. It's an iterative process – you will have to work through it several times to arrive at a result that meets the demands of both your pocket and your desires. All of these factors can remain open-ended as time goes on. I would never close the door, for example, on equity release mortgages or forget to review legacies or distributions if circumstances changed. It's *your* money, after all.

These factors are interconnected: the more valuable a home you want, the more capital you will need; and if the capital amount is already fixed, you will need to adjust your lifestyle accordingly. The more income you can still earn, the less capital will be required; the younger you are, the more income you are likely to be able to continue earning.

It's about tradeoffs – but then all of life is. No one gets to have everything that they want unless they are fabulously wealthy or choose to want very little. It's easy to see the tradeoff for example between a home and lifestyle: if you want a wonderful home, you may have to agree to travel less overseas, drive more modest vehicles or spend a great deal more time at home. If you want to put the emphasis on travel, entertainment and leisure activities, you may need to accept more modest accommodation.

Calculating these sorts of tradeoffs is relatively simple, as it is fairly easy to do the numbers. It is more difficult to be certain about how long the money needs to last (life expectancy) or to say in advance how much longer you would like, or be able, to earn additional income.

However, I think it's important to start thinking and talking about all of these factors as early as possible. You won't ever have it all worked out perfectly. But the alternative is paralysis and what a waste of life that is!

When I am doing wealth coaching work I always get the clients to calculate a 'Freedom Figure': how much they will need so that they can stop. Working out a Freedom Figure involves doing estimates for all the factors just outlined. Clients are often surprised to find that they already have enough to stop full-time work; others are far closer to their Freedom Figure than they thought. Follow the steps below to work out how much money you need to live the lifestyle you want.

TABLE 6: Freedom Figure calculations

1	Calculate your net worth (what you own, less what you owe).
2	Subtract the value of the home/holding in which you want to live when you leave the farm.
	Also subtract the value of any other major purchase you will make at that time, e.g. a boat, bach, new car, lifetime membership of a golf club, etc. You now know the amount of money that is available for investment.
3	Select the return (before tax) that you are likely to get from your investments.
	You will need to allow for inflation, especially if you do not have a heavy weighting of growth assets (which keep up with inflation). Calculate the income you would expect to receive annually from your investments.

| 4 | Add in any other income you expect to receive. |

You may, for example, expect to receive rent from the land or dividends from the farm or you may intend to have some paid work. You may also be eligible for income from superannuation. This should give you your total income before tax.

| 5 | Deduct tax to get an after-tax figure. |

| 6 | Calculate your expected annual expenditure. |

| 7 | Calculate the difference between your expected annual income and your expected expenditure. |

You will either have a surplus or a deficit. If you have a *surplus* then you are in a good position to get on with the lifestyle you want. If you have a *deficit* you should re-examine the numbers and the assumptions you have made. There are many places in which you can make adjustments:

- Is your budgeted expenditure for the future very high?
- Have you included all expected income?
- Can you earn some further income for a few years?
- Can you settle for a less expensive home?
- Can you do without some of the expensive purchases, e.g. a new car?
- Can you get higher returns on your investments?
- Are you being extremely conservative about estimating returns?
- Have you included all your assets in your net worth statement?

Once again, I tend to find that the clearer people are about what they want, the more flexible they are about juggling and adjusting these factors. If you are both really clear, for example, that you do want more time together with the family around, you will find it easier to lower your lifestyle expectations, such as annual overseas travel, in favour of having a nicer home in order to host lots of special family occasions and have the grandchildren to stay.

At the very least these factors open the discussion about what you need and want and how that translates into money. The earlier you start to plan, the more freedom you have to adjust the variables in your favour: it is much easier to earn income in later years if you have always planned for that by keeping up with current knowledge, acquiring new skills and developing networks that will offer work opportunities.

And it is all much easier if you have off-farm investments.

CHAPTER 8

Investing off-farm

Why invest off-farm?

Farms have traditionally been poor at providing income and profit but have concentrated capital value well. No matter how you look at it farming is a very risky business. So many things can derail any particular farm or farmer – not to mention the disasters that can befall the entire industry. It is extremely tempting to play 'double or quits' with your farm – the farm is always hungry for improvements and investment. It seems logical to keep reinvesting in the farm as you know it well, can see the opportunity it affords and are on hand to manage and protect your investment. However, all of your eggs are in one basket: there is no diversification at all. Farms are just as risky as all other family businesses and we know that many of them fail. The fact that the land and buildings are physical assets and continue to exist does not prevent disastrous loss of value through adverse weather events, disease or changes in legislation or the global market place. Your investment is highly vulnerable.

It's wise to make a commitment to begin the process of investing off-farm as early as possible. The amounts may be small initially; it's the principle (as opposed to the principal!) that counts and over time the funds allocated to investment beyond the farm can be increased. There are two main reasons to invest outside the farm:

Security
The off-farm investments are a hedge against anything going awry with the farm. I have noted already how risky farming and especially a single farm can be. This money is a form of security that can be used either to 'rescue' the farm or as a stake to begin again.

Succession
Over time the funds can provide capital for retirement so that the farm does not need to be sold to release capital or can be passed on to the next generation without too much debt being incurred. The off-farm investments are also a useful way of providing money for children who will not succeed to the farm. Farmers can use these other assets (property, businesses, another farm, shares, deposits) to assist siblings to set up in their own businesses or professions or to give them lump sums without burdening the farm. As I have noted before, managing these off-farm assets can be a very useful way of developing your children or even providing a career for one or more of them. Again, the timing of these gifts or disbursements will vary; but the process enables you to avoid the difficulties associated with putting too much debt on the farm in order to fund a retirement or pass money to siblings of the successor. Whether you intend to farm all your life and then sell or whether you wish to pass on the farm within the family it is wise to begin thinking about investing beyond the farm as early as possible.

Investment is very personal. There's no such thing as a 'good' investment – there's only investment that is right (or not right!) for *you*. What you should do depends a great deal on personal factors such as how old you are, how soon you anticipate ceasing fulltime farming, what other investments you have and how actively you wish to be involved in your investments. Every farmer will be different but the underlying reasons for investing off-farm remain the same:

- growing wealth;
- securing wealth;
- lowering risk through diversification;
- providing for retirement;
- making succession easier.

Investing beyond the farm is not hard to do and farmers make very good investors.

Farmers make great investors

Why do farmers succeed as investors? They are successful because doing well as an investor is not a matter of luck but rather of following the principles of investment. Farmers understand and work to sound principles in farming – analysis, buying well, taking care of their investments, understanding the value of timing and choosing to invest in ways that suit their place, position and personality. Farmers are also patient – you have to be, to farm well. It takes time to make investments work and you have to be prepared to stick with your investment plan through tough times as well as through the boom years. Farmers are conservative – in the best sense of the word – by nature and so take good care of their investments. Yet another advantage that farmers bring to investment is business literacy. All investments require an understanding of the numbers and an ability to evaluate the proposition. Farmers have spent years managing farm accounts, evaluating development proposals and dealing with bankers, lawyers and accountants and this is a great basis for venturing into off-farm investment. Investment is not about rolling the dice, going to the casino or betting on the horses. Some investors do well and others do not but that is not just a matter of chance. The ones who succeed do so because they follow the underlying *principles* of investment – it's not about luck!

Farmers also intuitively understand natural cycles as they have had long experience of the way these work on a farm. Good investors are also comfortable with the notion of cycles: they know that the drought always breaks; they understand that you have to survive the bad times in order to thrive in the good; and they know that the good times never last! You don't tend to hear farmers argue that 'this time it's different' or that the 'rules' no longer apply. There are natural cycles in investing just as on the land.

You do not need to become an expert in order to be a successful investor. You do not need an advanced degree in jargon and gobbledegook in order to make sound investments; but you do need to understand and follow the basic principles.

There are no silver bullets either – beware of anyone who promises you instant and effortless riches from your investments. Successful investment takes time and care just like everything else worthwhile – a successful farm, good health, relationships, children or achievement in any field.

There is some risk involved: to invest, you are going to have to give your money to someone else – even if only the bank. The alternative is to keep it under the mattress which is also risky. There is always a risk that you will not get your money back but that risk is small and manageable if you stick to the rules. Many of you will be nervous about deciding whom to trust – and rightly so. Anywhere there is money there are unscrupulous people looking to take it from you. However, if you stick to the principles and stay informed your risks are very low. Risk is something to be managed rather than avoided altogether – avoiding risk is an impossible goal in any area of life.

Remember that it's *your* money. You make the decisions and keep control. You can use other people to do much of the work for you but you do need to understand what is going on and you should never relinquish control. There is a big difference between delegating work and management tasks and abdicating involvement and responsibility. You should no more give away ultimate responsibility for your investments than you would for your children. (See chapter 15, 'Getting good advice', for more on advisers.)

Investment is fun. The more you understand the principles the more enjoyable it is. And it is not difficult. As you read the following chapters you will realise that you already know most of this – it's just common sense. Farmers make great investors – they have good life experience, are used to making pragmatic decisions and have had lots of practice at being sensible.

How to invest off-farm

I think of assets as being of two types: Wealth Creating Assets and Security Assets. I have written at length about these in *Get Rich, Stay Rich* (Allen and Unwin, 2003). Wealth Creating Assets are things that will make you wealthy. The benchmark used was that they grow your wealth by at least 15 per cent *per annum*. To get these returns you need a business, farm, highly geared property portfolio or actively managed and aggressive share portfolio. These assets have great potential to make you rich (as your farm hopefully also has!) but it is a really risky process – and takes all of your time, as you probably well know. Wealth Creating Assets suit younger people who are trying to develop wealth – they can take the risks associated with gearing because they have decades to make it work. Your successor probably recognises that as well – and may gear the farm and manage it aggressively for years in order to get capital growth. This is an undiversified approach and is therefore risky: all of your eggs are in one basket and you had better mind that basket well!

You have to take these risks to grow wealth whether it is through a farm, another business, property or shares. You may well spend many years of your life ploughing everything into Wealth Creating Assets: your own farm, a second or several other farms or even highly geared property or other businesses. This is appropriate and exciting and hopefully brings great satisfaction and creates great wealth. However it is not a suitable strategy as time passes and you get closer to putting in your last fencepost. How you choose to invest beyond the farm is likely to be very different depending on your individual risk profile – what risks you face at your age and stage.

INVESTING FOR SECURITY

Over time you should develop a second group of assets – Security Assets. By definition, these are assets that are less risky and give lower returns. Security Assets are things such as investment property (with low or no borrowings), a diversified portfolio of shares, bonds, cash deposits, a whole-of-life insurance policy and cash. Security Assets work to store and secure wealth: it will take problems across all sectors in all markets to knock you over if you have a good portfolio of Security Assets. This is rather different from having all of your wealth exposed to the next weather bomb or outbreak of disease.

Also you need to have some money offshore. Ask the 'foot and mouth question': what would become of your wealth if New Zealand got foot and mouth next week? Stock couldn't be sold, farm values would plummet and the fortunes of all related businesses and property would be jeopardised. Tourists would stay away. The value of the Kiwi dollar would plummet. Foot and mouth has occurred in Britain in recent years; Ireland narrowly escaped such a disaster. New Zealand is even more vulnerable from a wealth perspective as its economy is far more reliant on farming. In fact, it is very difficult to see how any business in New Zealand would survive even if it is not immediately obvious how it is connected to farming.

If the answer to the 'foot and mouth question' from the point of view of your wealth is horrendous then it indicates the need for you ultimately to have wealth secured off the farm, out of other farming businesses, away from your local community and preferably beyond New Zealand.

Securing wealth off the farm takes time. Clearly younger farmers will be unable to develop much of an investment portfolio because they won't have the free cash flow to do so. However it's the principle that counts and I believe that *some* Security Assets should be developed right from the get-go. The younger person will have little free cash and money to spare but could, for example, begin with a whole-of-life policy, and over time divert more money (and even more time) to off-farm investment as it becomes possible. As debt on the farm is paid off, however, it is important to look well beyond the farm for other investments. Farming is great for creating wealth but there are lots of other wonderful potential investments too.

Investing along a continuum

I see investing beyond the farm as a continuum: younger farmers concentrating on debt repayment and a little off-farm security; more established farmers exploring Wealth Creating investments, possibly using the farm for leverage; and older farmers moving to lock in wealth gains by building up Security Assets that can also be used to help with succession issues for both generations. This is an over-simplification: many younger farmers will not have great debt burdens; many older farmers will still be very active investors, and see themselves as asset managers managing a large portfolio of assets (of which the farm is only one) rather than primarily as farmers. There are obviously several factors at play, such as available cash, willingness to gear, interest in investment, time available and appetite for risk and each individual farmer's circumstances will vary. Working through the various scenarios is beyond the scope of this book but all farmers should be looking to invest off-farm in a manner that suits their situation and personality.

As you get closer to the second half of life you can make your portfolio more secure by paying down debt, reducing the number of Wealth Creating Assets and diversifying more. It is less usual nowadays for work to stop suddenly at a specific date. You may prefer to make much gentler transitions and blend aspects of full-time work with aspects of retirement. This means that the shaping of your portfolio will probably be more gradual too – unless you sell the farm outside the family, you may not have to make a decision about a huge sum on any one date.

Many farms however will be sold down progressively to a family member (see chapter 12, 'Prescription or discretion?'). This process can begin well before you are ready to cease full-time responsibility for the farm. Over time you can reduce your risk as you sell down further – less of your wealth is now exposed to the risk of farming. You can use these monies so released to diversify your wealth base and give you more Security Assets.

You can of course pursue the same strategy with other Wealth Creating Assets that you have invested in. For example, you might sell a highly geared property and use the proceeds to reduce debt on others. There is no need for these changes to be left until you are ready to retire fully; in fact, the right strategy is to have a plan to move assets in the direction of a secure, well-diversified portfolio over time. This process might take several years – it might be played out over your last decade in full-time farming – but there should be a plan. Plans are rarely followed to the date or to the letter; rather it's the underlying thinking and strategy that counts. Once you have decided on the kind of asset allocation you should have in place after the last fencepost then you can make appropriate choices as you go and when the opportunities arise in your portfolio.

What strategy will you follow?

It's important to have a strategic approach to investment. Without a thought-out strategy you are likely to invest (and disinvest) on a whim. This might be occasioned because you find yourself with some surplus; friends and neighbours are investing; or a particular investment opportunity is presented to you. This *ad hoc* behaviour is not a sound approach to investment and often results in a mixture of investments all over the place that in no way meet the investor's needs in terms of income or risk management. Some investment strategies to consider are:

Asset allocation

Making money through investing is largely about being in the right type of asset at the right time. Many people give this little thought: we are far more likely to try to pick the particular property or share to invest in, or to buy the 'right' business, than to consider the wider picture. Most of your investment returns are determined right from the start by the way your money is assigned to different classes of investment: asset allocation. In effect what this is saying is that what's important is the right *mix* of asset classes (ownership or growth assets such as business/shares, and property or debt assets such as bonds and bank deposits) at any time. For example, if the share market is doing really well in a bull run, those with lots of shares will tend to enjoy high returns, almost irrespective of the shares chosen; and *vice versa*. So the real choice an investor needs to make up front is what proportion of each asset class is the best fit for that person's age, objectives, other assets, etc. Once you have chosen this mix, the rest of the decisions are relatively minor. A Nobel Laureate has shown that over 90 per cent of your returns stem from the asset allocation decision. What this means in effect is that all of the effort that might go into picking individual shares, property or bonds is relatively unimportant to your returns compared to getting the asset allocation right. Trying to pick winning funds or shares is not worth it: the actively managed portfolios do little better than the passive ones that simply follow the market. This is great news for those of you who do not wish to be distracted from the farm while you are still working full time. It may also be music to the ears of those who have more time on their hands but who would rather select a book or tie a few flies and find a nice spot for the day!

Obviously the right asset allocation for each person varies. Some farmers will want to remain almost exclusively in full-time farming with some money invested beyond the farm, probably passively managed. Others may wish to develop their portfolio of investments so that over time they will be a very active manager of a range of assets, one of which may be a farm. A younger farmer who wishes to invest aggressively beyond the farm (probably using the farm for leverage) and who wants to become an asset manager might well build a property portfolio or a carefully chosen and actively managed share portfolio. An older farmer who is looking to lower risk by diverting money off-farm might have a well diversified portfolio of shares, property and deposits that are in managed funds. You might fit anywhere on this continuum depending on what you have already accumulated, and on your age, stage, debt, time available, skill, interest and personality.

Buy and hold

This strategy works best the more time you have on your side as over time the markets trend upwards. However, there can be long periods where they go nowhere or even lose money. The key idea here is that you are in the markets for the longer term and expect to get good returns from the overall growth of the market over time.

I think there's a lot to be said for 'set and forget'. I don't mean ignore – you will need to do the occasional review and reallocation of your money. However, unless you are the type who wakes up in the morning and has your portfolio set up as the default on your computer, and who then reads the business pages of all of the world's best newspapers and blogs online, set and forget is for you. Let the markets do their work!

In and out

This requires you to be a contrarian and to buy when there is a collapse and sell out early when the markets start to boom. It is easy to describe and difficult to do – everyone else will be doing the opposite, after all. Timing this correctly is not easy either – after all, how much further might the markets still drop and shouldn't you wait longer to benefit from the rising markets? However, it gets easier with practice; especially if you can discipline yourself to stop looking for the absolute bottom to buy and be prepared to leave some money on the table in the rising markets. After all, you just need to make good returns, not get every last cent. Following this strategy allows you to do well even if the markets are essentially going nowhere over a decade or two. Various people are credited with saying that they made a fortune by selling too early. There's real wisdom in that: getting out while there is still money to be made but making sure that you are out before it's too late!

Whether you have invested much off-farm or not you will need to reorganise your wealth as you approach your last fencepost…

CHAPTER 9

Reallocating your wealth

The first half of our lives is all about creating wealth. You will have worked hard to make the farm profitable and more valuable over time. Almost everyone has to borrow to create such wealth (whether for the farm or off-farm investments) and live with the risks attached to that gearing. However, there comes a time when the game is no longer about growing wealth but rather using that wealth to live the life that you want. Otherwise what is the point?

Therefore the second half of life is about using wealth. This is a major change for many to take on board: when you have spent all your adult life to date trying to make the farming business flourish and to grow your wealth, it can take a big attitude adjustment to consider doing anything else. But there comes a time when the job is largely done. You may no longer be able to work as hard or devote as much energy to the business as before – and you may not want to! Likewise, it is unwise to continue to grow wealth through leverage – I have already mentioned many of the risks associated with farming and gearing only amplifies these and the risks attached to any other investments you may have.

Finances beyond the last fencepost are all about using capital for income. Most of your previous approaches to investment have to change; and all of the assumptions you have been making (e.g. about gearing and levels of return) will need to be revisited, as they are mostly inappropriate for this stage. You now need to have much less risk in your portfolio. That almost certainly rules out gearing or leverage and means that if you have debt you should probably look to discharge it or at least reduce your borrowings. You will also need much more reliable income from your investments at this stage as you will probably be mostly living off capital. As you will see, this does not mean that you must be solely invested in cash or bonds but does mean that you must think carefully about the income flows that you need and plan carefully to ensure that you have cash as you need it. When you were in the wealth creation phase you would have been focusing mainly on growth; this focus will change, though some growth is still necessary especially if you have decades of life expectancy. It is time to take action.

Change your attitude

The biggest change is getting your head around the idea of consuming some of your capital rather than focusing on simply growing wealth. This, after all, is what you were working hard for all your life – and the time is here! Most investors are focused on asset accumulation. Thinking about using some of your capital for income is a fundamental shift because you have probably spent a lifetime growing it. You now need to focus on asset preservation and distribution. Income will largely come from capital rather than work.

Reduce risk

You had to take on risk to create wealth. Farming is very high risk as I keep pointing out but presumably if you are reading this it has paid off in your case. However, many of these risky habits now need to change – you cannot afford the risks of losing much of your capital at this stage in life. Think of it like this: making more money now isn't likely to change your lifestyle, but losing it will! You may already have achieved some diversification by investing beyond the farm. You should probably diversify further in order to reduce your risk: spread your assets even wider. It's far too risky to have all of your wealth in one asset class – business/farming, property and shares. You will need to have some assets in bonds and deposits so that you have some debt assets as well as ownership assets. The rationale for this is that the performance of different asset classes does not usually correlate: when one is doing poorly, the other (generally) is doing well. This fact may require you to sell some assets in order to diversify. That is also particularly true if you have all or most of your money in New Zealand – a large proportion of it should now be offshore. Holding all of your wealth in New Zealand is too risky a strategy at this stage of life: the New Zealand economy is small and fragile and it is easy to think of single events any of which could wipe out most of that wealth. You can't afford to make big mistakes in terms of where you put your money at this stage of life as you do not have time to recover the wealth. Your aims now differ from the time of your life when your focus was on the creation of wealth, and your asset allocation should reflect this change.

Clear or reduce debt

Gearing has probably contributed a great deal to making you wealthy. Most farmers would find it impossible to acquire and develop a valuable farm without borrowing. Leverage allows you to turn the wheel faster and amplifies investment returns and you may even have geared some of your off-farm investments. It can be hard to break the habit especially if it has served you well and given you high returns on your investments. However this is an inappropriate investment strategy beyond the last fencepost: you will have to get used to ungeared returns. Borrowing is simply too risky at this stage: it gears your losses as well. At the very least, it is debt that forces the investor to sell at an unfavourable time. I would recommend that you carry no debt beyond the 'retirement' point or a small amount at the very least. This may well be the time to sell some leveraged investments, clear the borrowings on others and make sure you are well diversified with the remainder.

Review your asset allocation

It is typical for farming folk to have most of their wealth tied up in the farm. Even if they have diversified off-farm it is likely that they have invested in allied businesses or properties. Many farmers, for example, help finance a son or daughter into another farm – the same type of business and often in the same locality. They may buy *local* residential, industrial or commercial property. Farmers are also attracted to other agribusiness – pharmaceuticals, fertiliser, farm machinery, bioengineering, horticulture, etc. – because they are interested in these industries and have more than average knowledge and insight. The problem with this sort of investment – understandable as it is – is that you are prone to a 'domino' effect: anything that would severely damage one of the investments such as an outbreak of disease is likely to have a knock-on effect on many or all of the investments. Now is the time to take a really good look at your portfolio and think about it from a different perspective. If this portfolio has to provide reliable returns over the coming decades it needs to be diversified beyond agribusiness; and it must be diversified geographically to withstand local shocks or poor performance in this region of the world by international standards. Your portfolio should also be well diversified across asset classes so that you get both the benefits of some growth from shares and property and also the greater certainty of returns from interest-bearing deposits and cash. In the past investment advisers put the money of people in the second half of life mostly in cash and bonds – safe as the bank, but with no growth. This made sense when you didn't expect to live for much more than a decade after 'retirement' but that is no longer the case. You will still need some growth in your portfolio but this will need to be organised in such a way as to reduce your risk. You are probably better to own property through REITs (Real Estate Investment Trusts) that allow you to invest in many properties in various locations rather than directly investing in a small number of properties. Similarly your shares should be well diversified. You are now seeking a share of the overall growth that can be expected in share and property markets over the coming decades, rather than trying aggressively to create wealth by carefully picking a few shares or properties.

It's really important to get your asset allocation right as this is what determines your returns. The asset allocation that's right for you will depend on several factors such as your age, how much income you will have from other sources (such as part-time work and New Zealand superannuation) and how much time you expect to spend overseas. The longer your life expectancy, for example, the more you will want allocated to growth (shares and property) in order to keep up with inflation; the more time you expect to spend offshore (maybe visiting family), the more foreign currency you would want to hold to hedge the New Zealand dollar. It's very difficult to be specific as clearly each person's circumstances vary but, generally speaking, I would suggest that those close to putting in their last fencepost would have 40 to 60 per cent ownership/growth assets (shares and property), and 40 to 60 per cent debt assets (bonds and cash deposits). I would suggest that 40 to 60 per cent of these assets should be offshore. That is a very big shift from having 100 per cent of your wealth tied up in the farm. It's also quite a departure from traditional advice to those entering retirement, which was to have 100 per cent in debt assets; but, as I have pointed out, you are unlikely to be fully 'retiring' and, even if you never do another paid day's work, your second half will be nothing like that of your parents.

Dollar cost average

Many aspects of investment are important to understand but are beyond the scope of this book. However, I have to make space to emphasise dollar cost averaging, as I have seen the flouting of this principle cost so many people, especially farmers, so much. The basic premise is that it is impossible to time markets accurately either for entry or exit – by the time you know that it's a good time for either you are too late! Obviously younger people can, for example, reason that markets are in a low and it's therefore a good time to invest. If they get it a bit wrong and markets continue to fall that's regrettable, but not likely to be terminal – as they have decades of investment time ahead. However, you cannot risk these unexpected losses as you approach succession, or afterwards, as you do not have the recovery time. Older investors cannot risk trying to time markets. You may not feel that you are trying to time the market but that is in effect what you are doing if you place or withdraw a lot of money at one time. Farmers, especially as they approach their last fencepost, need to reallocate their money gradually. Think in terms of making a *gradual* transition from one asset allocation (probably mostly in a farm in New Zealand) to another (probably a well-diversified portfolio with a good deal offshore).

The answer to this is to dollar cost average. Dollar cost averaging works by taking the timing element away. You either 'drip feed' your money into the markets or you withdraw it gradually. This means that you get an *average* of prices over a period of time. It means that you spread the money you have for investment by making smaller investments over several years rather than investing all of it at one time. This is especially pertinent for farmers who may have very large amounts to invest: you have sold the farm, or your child has made a significant repayment of the debt owed or you have just received the annual rental from the farmland that is leased to your child. You may intend that all of this income be invested. However, how do you pick the day? Is today a good day? Will tomorrow be better or worse? What about next week or month? I have seen situations where despite good returns the original amount invested from the proceeds of a farm sale had not recovered the original value five years later, because all the money was invested on the same day – and it wasn't a good one. The financial adviser had done good work, the asset allocation was appropriate, everything was in order – but the money was all invested at the same time, and the time was not propitious. I, like you, am usually anxious to outperform the markets but you can't take that risk with your money as you get close to the last fencepost.

Now is also the time to look for more income.

CHAPTER 10

Income after your last fencepost

One of the biggest hurdles facing many families as they consider succession is to solve the income problem for the older couple. You can frame this as widely or narrowly as you wish: some farmers want a lump sum from the farm that will allow independence for the second half, while others may only want enough for another house initially but need a stream of income to be maintained from the farm.

Living off capital

As you will have spent most of your life building wealth you will have been much more focused on capital growth than income; and you presumably had income from the farm. However, if you are now to live off capital (even if you continue to do some paid work) you will be more focused on getting some income from your investments. It is a mistake simply to convert your other assets to cash deposits and bonds though the income from these is fairly predictable. The problem is that there is no growth and you still need some: after all, you may live for another forty or fifty years. Putting in your last fencepost does not mean that you are old – you may well be only halfway there! You will need income but you will also need some growth: it's possible to get both.

Some of the things you may want to consider include:

Selling the farm
This strategy works well either when you do not have a successor or when taking enough money out of the farm – loading it up with debt – is unsuitable. Even if you have a child who wants to farm, it is often easier to cash up and help the child into another, more suitable, farm. This releases a lump sum for investment to produce income.

Gearing up the farm
Another way of getting money out is to allow your successor to load up the farm with debt so that you can withdraw some capital. Younger people are in a position to carry and service this debt while you need to secure your capital by taking some of it away from the farm. Many farmers are reluctant to so encumber the younger generation but I believe it is a very appropriate strategy and is fair to both parties. Again, this money can be invested to provide income.

Retaining the land but not the business
It is relatively simple to pass the management and ownership of the farming business on to the next generation but to retain ownership of the land, probably in a family trust. The business will be leasing the land from the trust and this rental can provide an income stream for the succeeding older couple. The risk with this strategy is that your future income is likely to be highly dependent on the

fortunes of the farm and farming in general. There is little or no diversification here until you use much of this income to invest in other assets.

How do you get income?

You need to work out what kind of annual income you will need or want over the coming years (as discussed in chapter 7, 'The numbers game'). The money that you get from the farm needs to be invested in such a way as to reliably provide this income. Financial advisers used to recommend that all of this money should be invested in debt assets – bonds and cash deposits – to give assured income in the form of regular interest payments. This makes sense up to a point because people want assured income. However, these asset classes have no growth so inflation is a very real problem unless you are already so old that you can afford to ignore it! Many farmers will have several decades left after they put in their last fencepost, and as discussed will need to allocate a good deal (40 to 60 per cent) of capital to ownership assets (shares and property) initially – in order to benefit from the ongoing growth of those asset classes. Obviously, the proportions are likely to change as you get older: people in their eighties do not usually need half of their money to be in growth assets. This is what gives you a hedge against inflation: as inflation increases, so too does the value of these assets – thereby future-proofing your income. When you are getting your income from capital rather than work you must be very careful to inflation-proof that capital.

Likewise you must take care not to lose that capital, which is why a good deal of it needs to come out of the farm and be spread widely across a range of asset classes – with much of it offshore to insulate it from the vagaries of the New Zealand economy. None of this is difficult to do in practice once you accept the need to do it. It is a big mind-shift – after all, you have spent a lifetime growing your capital aggressively by putting almost all your time, capital and returns into a single asset. Hopefully, that strategy won big for you, but it is the wrong strategy for the second half; and it's a strategy for wealth creation, rather than one for using capital for income. Obviously, you should have independent financial advice to help you restructure in this way.

Getting an assured and smooth stream of income can be a challenge when you are living off the returns from your investments and do not have everything invested in bonds or deposits. You may have made an effort to invest in high-income shares but they may or may not provide the expected dividends – and dividends from some may stop altogether! Your property holdings are likely to be through listed property trusts (REITs). The returns will go up and down and these investments are volatile. This state of affairs can be unnerving when you are depending on the income for living costs and wishing to be sure that you can take a trip, buy the grandchildren some presents and pay your bills.

There are a few simple approaches to help smooth out your income:

Keep a pool of cash

If you are the kind of person who will worry about when the next dividend or interest cheque will arrive then it's a good idea to keep plenty of cash. Depending on your personality, you might hold anything from six months to two years of living expenses in the coffers in cash or term deposits. You will still earn interest and can stagger the maturity dates of the term deposits to get the best terms – but without locking up your money for too long. This means you don't need to be anxious in the shorter term about what you spend as long as it's within your overall budget, and the coffers will keep being replenished from your investments. A pool of cash provides psychological insulation from what is happening in the markets.

Keep separate 'buckets' of money

It is perfectly reasonable to divide up your money in terms of different purposes. You might for example keep quite a bit in cash for the coming year or two, and also have more in bonds for a few years out. You can invest more aggressively with money that you do not expect to access for a decade or more. This would be true for example of funds you have earmarked for legacy purposes – the risk profiles of those for whom you are holding them are very different from your own. Depending on how much money you have and your interest in investing you may want to set aside some money for 'playing' with appealing investments. The idea behind the bucket strategy is the organisation of your investment money according to purpose.

Be prepared to sell some assets or units from time to time

Some of the better performing shares or funds may not pay dividends. In fact, higher dividends are more common in New Zealand than in other countries. You may see the value of these assets appreciate but still get no income from them. You can solve this problem by selling off some shares every now and then and banking that money to pay living expenses. This process does not change the amount you have invested in these shares; cashing up some units simply releases the gains you have made.

Keep rebalancing your portfolio

Because of the vagaries of the markets – and you may be in these markets for thirty to forty years – your investment portfolio will get out of balance from time to time. Take, for example, a couple with 50 per cent in ownership assets and 50 per cent in debt assets. In boom times the ownership assets (business/shares and property) are likely to gain a lot of value and soon may be 60 per cent or more of the value of the overall portfolio. Good you say? Well, yes, in the sense that you have enjoyed good performance and high returns.

But, if the original fifty–fifty allocation were correct, it's time to sell some of these assets and buy some more debt assets to rebalance. Remember that your assets were allocated to give you the best returns for the level of risk that you should take and it's important to maintain that allocation unless something significant has changed. Remember, too, that if you still have a significant financial interest in the farm, this fact needs to be considered for rebalancing – if the farm has gained greatly in value, this may be unbalancing your total asset allocation.

Now that you know what you want and can think about how to use your wealth for income let's look at the practicalities and technicalities of transferring the management and/or ownership to others.

Chapter 11

Wills, trusts and succession

When you are establishing your formal succession plan, there are some issues that you will have to resolve. Some of these issues may not be able to be resolved completely – there may be matters that conflict with each other. I often hear, for example, that farmers want one of their children to have the farm without debt (the family has worked for thirty years to clear the debt) but also want all the children to be treated equally. Well, in most cases, these aims cannot both be met – the very best succession plan cannot make things out of thin air! Compromises will need to be made and aims prioritised.

There does come a time when your succession plans have to be documented. Hopefully, these plans have emerged naturally through a series of family meetings and everybody knows what is going to happen. The formal documentation will be through your will, and probably through the family trust and a letter of your wishes. Before things are documented, and those documents signed, you need to have thought of a way of resolving as many of the issues as you can. This all needs to be done as early as possible – you never know when your will is going to have to be read! Even people just starting out owning a farm and/or starting to have children need to have a succession plan that will work for the family. You may only be thirty-five, and the twins may not yet be two, but you need to be sure that they can take over the farm or the proceeds from the sale of the farm. It is a horrible thought, but it just may be that neither you nor your spouse will be around to watch over the children as they grow up and the succession plans that you put in place now may not be able to be changed in the future.

There are some issues that you need to confront.

Most farmers have only one big asset: the farm

Sometimes the big asset, the farm, is able to be split but mostly it is not. At the same time some of your children may want to farm, others won't. It will be difficult indeed for all the children to receive equal amounts when you are gone. Unless the family is prepared to increase borrowings, there is not going to be equality; this does not mean that there cannot be fairness. However, if one of the children wants to carry on with the farm, the others are unlikely to receive as much when you are gone. This issue will be a main agenda item when you have your family meetings. It is also something that needs to be set out when you document your succession plans through your will or the letter of wishes that goes to the trustees of your family trust.

You have to be ready for succession to happen at any time

I have some bad news – you will not live forever! One day the farm will pass to someone else. Succession may happen long before you would normally expect it to and while the children are very young. If you are in your thirties or forties, it is unlikely that you are going to die soon – but it does happen. Moreover, if you do die, it is quite likely that your partner will die with you (in a car crash, house fire etc.). Therefore you need to have thought about what succession will be like if both of you go prematurely. Your children may only be five years old and at that age no one will know what they are going to want to do in life (and will not know for at least another twenty years). A plan that is flexible and gives a great deal of discretion to those who will look after your children's money is important.

When you put in that last fencepost and walk off the farm, you will need income

If one of your children is to take over the farm, that child will need to buy it from you. This in turn means that the child will have to borrow – something that the family may not find very attractive (after all, you have probably worked for thirty years to get the bank paid off). However, if the child taking over the farm does not buy it from you, you may not have any money with which to buy a house in town, or to invest in order to generate income on which to live. Furthermore, if the child is given the farm, rather than buying it from you, the issue of fairness probably arises again. It could be that your son or daughter who is taking over the farm pays you a regular amount so that you have income on which to live. However, this means that your lifestyle once you have left the farm will continue to be completely reliant on the performance of the farm. That may have been reasonable when you were young; however, when you are retired it is very important that you have a diversified set of investments. These are issues which many farming families have to weigh up – there is no magic answer that serves everyone. The solution is peculiar to each individual family.

Your children will have relationships

It may be quite a way off but it is probable that at some time in the future your children will have relationships. Anyone who has children needs to be thinking about the partners that their children may have. As discussed, this applies just as much (if not more) to those with young children: it is entirely possible that a partner of one of your children will end up with a big part of what you have built up – and you may never meet this person. Any succession plan needs to take this into account. It may be that you are quite happy for your children's partners to take a part of the bequest that you leave, particularly if it has been a relationship of long standing and there are children involved. However, the Property (Relationships) Act means that any relationship that lasts for longer than three years can see all assets split equally. With many people having live-in relationships when they are young, you do need to be very careful that substantial amounts of money or other assets do not go to your children when they may still be vulnerable. Again, these are principles regarding which you might have agreed to policies at family meetings as your children grew up. One way or another you will need to decide on and document your wishes.

There may be other issues that you face – issues that are peculiar to your own family circumstances. For example, you may:

- be farming in partnership with a sibling or another family member;
- have a child who is disabled and who will need additional long-term care;

- have a farm owned by a family trust that was set up decades ago and does not have the flexibility of modern trusts;
- have four children who wish to farm and want help to get started.

While there are solutions to many of the issues that face farmers when they are thinking of succession matters, sometimes not all issues can be resolved. A compromise is likely to be needed on the issues, and a decision made about which are the important questions that you really want resolved, and which are of lower priority. For example, it may be that your son simply cannot be given the farm if you want a reasonable lifestyle in retirement – your son will have to buy it from you, loading up on debt. Family meetings, as before, are the best place to explain and discuss these choices. You will need to decide what is most important to you, giving way on some issues to resolve others for the best possible outcome. The final package that you come up with will probably not be perfect but will be the one that satisfies your highest priority.

CHAPTER 12

Prescription or discretion?

Fundamentally, there are two ways of documenting your succession wishes: you can be definitive or prescriptive about what you want to happen or you can take a more open-ended or discretionary approach.

A PRESCRIPTIVE APPROACH

Taking a prescriptive approach means that you lay down exactly what you want to happen and allow no variation from your plans. This is almost always done through your will which will be drafted in such a way that the trustees are bound to carry out your instructions. Following the prescriptive model of succession will mean that you will have to decide at the time that your will is drafted who is to get what. You will have to determine which of your children will get the farm and what the others will get. If you are like many farmers, who own very little other than the farm, the result may be that on your death one of your children will get the 2,500 hectares along with all the stock and plant, while the other will get the collection of teaspoons that you accumulated on your trip to Europe in 1962. With the prescriptive approach there may be fairness issues!

Moreover, prescribing exactly who will get what does not take into account any changes that occur between the documentation of your wishes and when those wishes are executed. There may be changes within your family (one child decides the farm is no longer desirable or another becomes seriously disabled and needs a lot of money for care). There may also be changes to market values (farm prices collapse and at the same time the teaspoon collection turns out to contain a spoon that Julius Caesar used to stir his tea and which is now conservatively valued at $23 million). Prescribing exactly what you want to happen means that your executors can take no account of changes in circumstance. And you must be mindful that succession can happen at any time leaving you with no opportunity to update your will.

A discretionary approach

The discretionary approach is organised through a family trust. Family trusts are discretionary by nature – the trustees have complete discretion as to which beneficiaries will benefit from the trust, when they will benefit and what benefit they will receive (see chapter 13, 'Family trust basics'). Distributions from trusts do not have to be made at any particular time and trustees can drip feed assets to beneficiaries when appropriate. Taking the discretionary approach means that nothing is laid down as definite: changes can be implemented so that distributions are not made unless it remains the right thing to do. It is true that a prescriptive approach can also be changed – you can change your will at any time. However, if you die before you have updated your will to better deal with changed circumstances, the existing will applies.

The problem with the discretionary approach is that you have to assume that you will not be around to ensure that the trust distributes to the beneficiaries as you would want it to. When thinking about passing assets on to the next generation, we have to assume that the assets are passing on because we have passed on also. When you have died, new trustees will be appointed and it will be these people who exercise their discretion – they have to use their good judgment. In this respect discretion is a double-edged sword: it is good to be able to keep your options open; however, it will probably not be you who chooses which option to take. There is much that you can do to ensure that the new trustees manage the trust and distribute to trustees as you would wish them to. In particular, you must choose very good replacement trustees and give them guidance through a letter of wishes (see chapter 13, 'Family trust basics'). Nevertheless, if you follow the discretionary approach to succession planning, there is always a chance that things will not be managed as you would wish.

Discretionary or prescriptive approach?

On the whole I prefer farmers to adopt the discretionary approach to succession planning: this will mean using a family trust. In particular I prefer the discretionary approach for younger farmers. Farmers in their thirties and forties are likely to have children who are young and who have not yet made up their minds about whether they want to farm or not. (In fact, they may be so young that they barely have minds to make up!) These younger farmers have to keep all their options open. They also have to be mindful that there is some chance that they will not be around to watch their children grow up. While there is not a very high chance that people in their thirties or forties will die at that age, it does happen – and so succession plans need to be made with that possibility in mind.

Furthermore, sadly, there is a chance that both you and your spouse will die together (in a house fire, car or plane crash, boating accident, etc.). Plans have to be made that will be robust and will work well if neither of you are there to take care of your succession arrangements. It seems obvious to me that younger farmers should have a discretionary form of succession arrangement – it is impossible to predict all the different circumstances that may arise. One of your children may become seriously ill, or one of them may want help to buy a business. It could be that one of your children becomes a semi-permanent student and wants to continue to live off money from your estate or trust, while another would like to get onto a farm. There is no way that you can predict any of the things that may happen within your family after you have gone – especially if your children are young when you go.

As you and your children get older, the discretionary method of succession generally becomes less suitable. As you age you can afford to be more prescriptive: your children's paths are more set, and it is probable that a child who is going to take over the farm is already managing it and possibly even owns it to some degree. When you are in your sixties, the children are likely to be older and by that time you would hope that what is to happen to the farm will have been settled.

Many people also use their family trust to distribute assets to their children in their own lifetimes. For many, succession does not happen at the time of death but instead happens progressively over a decade or two. This means that part of the ownership of the farm can go to one of the children (with other assets possibly going to others). More commonly, perhaps, you may (via the family trust) help the children into business by way of providing borrowing guarantees. Sometimes the business may be a farm or small block of land that one of the children can farm. What the children want or need should emerge over years of development, conversations and family meetings. It's relatively easy to see how you can meet their individual needs and interests if you begin early.

Family trusts are extremely flexible ownership entities – they allow the distribution of both capital and income to any of the beneficiaries at any time. No reason is needed for making the distribution – it is completely at the discretion of the trustees. Also the trustees, if they choose to do so, can have the trust give a guarantee in favour of a beneficiary, which could allow a child to buy a starter farm. Furthermore, the trust does not have to distribute all the assets of a trust at one time. The trust can drip feed assets to beneficiaries over time. Some farms are owned by a company, the shares of which company are owned by the trust. The trustees can, if they wish, distribute some of the shares to a beneficiary, holding onto others for as long as they think appropriate. This flexibility is a major benefit of ownership through a family trust.

Relationship property

One of the issues that farmers face with succession is their children's relationships. The Property (Relationships) Act means that there is generally a sharing of assets when a live-in relationship that has lasted longer than three years fails. Anyone with children, especially those with young children, needs to be aware of this. If you prescribed your succession plans in your will, and died while the children were still very young, the children would receive their inheritances at age twenty, or when the will prescribed, without any check on the nature or state of the children's relationships. When the children receive their inheritances, they may already be in some kind of live-in relationship or could go into one soon after. If they use what they have received from you within the relationship, half of the inheritance could belong to the partner (a partner whom you have probably never met).

If, however, the succession arrangements are managed by a discretionary family trust, the replacement trustees can exercise their discretion before they make any distribution. This would mean that they would contact the beneficiary and before making the distribution have a chat about any relationship. It may be difficult to be sure about the state of a relationship. However the trustees would usually act cautiously and decline to make a distribution if they thought that a relationship was not sound. This would be particularly so if the children were in their teens or twenties (it is not at all unusual for people of this age to be living together).

Fundamentals of the discretionary approach

The discretionary approach to succession is mostly suitable for younger farming families. Generally, these farmers are not yet sure which of their children might want to farm and what they will want to do themselves in their retirement. In fact, it is often the case that the children's plans and their own really only become firm a few years before they put in their last fencepost. Up until plans become firm, it is important to keep all options open – and that means taking a discretionary approach to succession.

When we at Wealthcoaches are advising people on succession, we have a basic plan that we use for most clients. This plans works for farming families just as well as it does for other people who are in business or who have substantial assets. The plan is especially useful for those who have children who are not yet at the stage when they can be 'safely' given large bequests or when they have not yet finalised their future direction (i.e. they are not sure if they want to farm or not). It works on the assumption that the children should not receive major inheritances at too young an age; they should learn to make their own way in the world, and develop themselves professionally, rather than waiting around until they reach a certain age to get a big handout. The children may receive help (for education or to get into some kind of farm or business), but they will not receive big amounts of money that they can spend however they like until those looking after their money for them are confident that they are ready. The plan also works on the assumption that between the time of your death and the time that the children are ready to receive major distributions anything can happen. If you die when the children are, say, ten or twelve, they could become ill or disabled, want to do a PhD in microbiology at Harvard, get married and divorced, start a successful business, be put into a Turkish jail, try out as a professional rugby player or start an unsuccessful business. The discretionary approach means that trustees can give them the right amount of help at the right time.

The basic model is:

Your will leaves everything to the family trust
This includes all major assets (such as the farm, if it is not already owned by the trust, and other investments). It also includes any debt that the trust owes you that has not yet been gifted to the trust. Only personal assets (such as clothes, books, jewellery, heirlooms) remain outside the family trust and your will may dictate where each of these assets goes.

Your will appoints guardians for the children if they are of an age to need them
The guardians should not usually be the same people as those who are acting as trustees – the guardians should be required to go to the trustees and request funds for the children as they need them.

Your will appoints new trustees and/or a new appointor of the family trust
These people will be carefully selected by you for their expertise and knowledge of the family.

A letter of wishes is drafted
This gives guidance to the replacement trustees as to how you would like the trust to manage the assets and when and under what circumstances distributions are to be made to the beneficiaries. This letter does not bind the replacement trustees but it is most likely that the trustees will follow your written wishes provided that circumstances do not change radically (e.g. one of your children keeps on ending up in that Turkish jail). Because of the discretionary nature of the family trust that owns the assets, the letter of wishes is one of the most important documents in your entire succession plan.

In lots of ways this postpones many of the decisions as to what will take place and how your farm and other assets will be dealt with. However, it is better for younger people with children who are not fully grown that these decisions are postponed – you do not know how things will pan out and it is better that you outline what you would like to happen (via your letter of wishes) but leave the firm decisions to trustees who will know the situation in detail at the time. Clearly, this means that you place a great deal of faith in the trustees – these people will effectively be standing in your shoes, hopefully making the same decisions that you would have made if you had been here to make them.

This is a basic outline of the model that we usually adopt when it is appropriate to have a lot of discretion in the succession plan. Subsequent chapters deal with some of the issues in more detail.

Chapter 13

Family trust basics

Most farmers own their farms through a family trust. Usually, the land and buildings will be owned by a family trust with the farming business, including the stock and plant, being owned by another entity (perhaps a company or partnership). Most farmers are therefore familiar with family trusts and this chapter is only a basic reminder of some of the concepts.

Components of a family trust

A trust has a number of parties and features:

Settlor
The person (or persons) who establishes the trust is called the settlor because this person 'settles' an asset on the trust.

Trustees

The people who own the trust's assets and make all the decisions regarding the trust are the trustees. In particular, the trustees decide which, and when, beneficiaries will receive distributions from the trust. I have said before that family trusts are discretionary, which means that trustees are unrestricted as to who will benefit and by how much from the trust. This discretion lies completely with the trustees. When a family trust is formed, the settlor is usually a trustee which gives you, as settlor, effective control over the trust. However, you have to make provision for new trustees when you are gone and clearly a great deal of thought needs to go into this.

Beneficiaries

The beneficiaries are the people who will receive distributions from the trust. You are likely to be beneficiaries of the trust along with your children and grandchildren. Sometimes wider family members or even close family friends may be beneficiaries though all beneficiaries must be due 'natural love and affection'. No beneficiary has any right to receive a benefit under the trust – they have the right to be considered but cannot sue for or demand a benefit.

Appointor

The appointor features in many modern family trusts. The appointor (it is usually the settlor(s) at the beginning) can remove and appoint trustees and quite often can remove and insert new beneficiaries. Given the ability to hire and fire trustees, this person has the most powerful position on the trust. A new appointor will usually be named in your will and, obviously enough, naming a replacement appointor will be the most important decision in your whole succession plan.

Trust deed

The document that is executed to initiate the trust is known as the trust deed. The trust deed is the trust's 'bible' and its provisions must be followed. You would need to go to the High Court to change a trust deed and this is neither easy nor inexpensive. Fortunately, most trust deeds are drafted to be very flexible allowing for many changes over the life of the trust (e.g. most allow for the appointment of new beneficiaries); although you occasionally do see trusts that have deeds with special provisions or without clauses that allow for future change.

Establishing a trust

A trust is established when the settlor(s) passes over an asset (usually only $10) and the deed is signed. This gets the trust going; although you will, of course, want more valuable assets to go into the trust later. Diagrammatically, the formation of the trust looks like this:

Figure 1: Formation of a trust

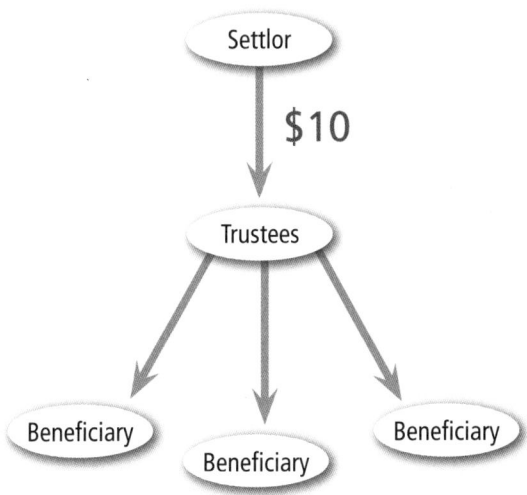

Family trust basics • 119

At the same time that the trust is formed, the farm (and/or other assets) will be sold into the trust. This is done by a sale and purchase agreement for the sale of the farm. The purchase of the farm by the trust is funded by way of vendor finance: you, as the vendor of the farm, finance the trust's purchase 100 per cent (the trust has no money except for the $10 that you settled on it). An acknowledgement of debt is also drawn up whereby the trust acknowledges the debt that it has to you for the purchase of the farm.

Diagrammatically, the formation of the trust and the sale and financing of the farm looks like this:

FIGURE 2: Sale and financing of the farm through a trust

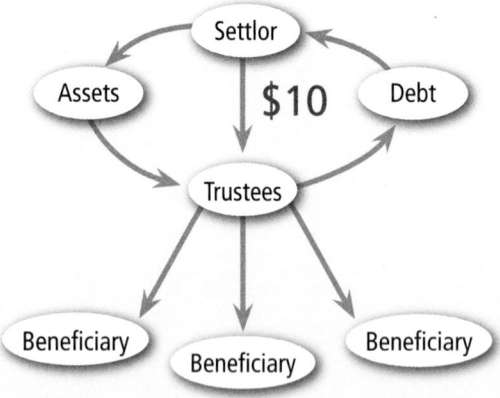

GIFTING

Once this transaction is completed, you no longer own the farm – the trust does. The thing that you own is a farm-sized debt, i.e. the trust owes you an amount equivalent to the value of the farm. You will make gifts to the trust by way of forgiveness of this debt. In New Zealand you can gift $27,000 (or $54,000 for a couple) *per annum* before attracting gift duty and most farmers gift the maximum that they can each year.

With the current value of farms, it is unlikely that you will be able to gift the entire amount of the debt in your lifetime; for example, it would take a farming couple thirty-six years to gift a $2 million-farm. Therefore, most people forming a trust change their wills so that on their deaths any remaining debt to the trust is forgiven. The forgiveness of debt that happens on your death does not attract gift duty (although I do think that dying is a rather extreme way of avoiding a government tax!)

When the trust is formed and the farm is sold into it, you will probably want to change your will so that other assets that you own go to the trust on your death. This would mean that on your death all the valuable assets that you own will belong to the trust and the trustees can deal with them according to the deed, your letter of wishes and the circumstances at that time.

Trustees

It is very common for the settlors of trusts to be the trustees and appointors. This gives you effective control of the trust in your lifetime. I recommend that you also have a professional (lawyer, accountant or trust company) as a trustee. Four reasons for having a professional as one of the trustees are:

1. The professional will give the trust continuity after you have gone.
2. The professional will make sure that the trust is well managed and administered.
3. The professional will keep all records safely and keep a diary of when gifting is to take place.
4. The professional will keep abreast of trust and tax law and advise you when there are changes.

Most farming families have either their lawyer or accountant as the professional trustee. I recommend that you choose the one to whom you feel closest. Remember that this person is likely to continue in the role after you have both gone and therefore you want someone who knows the family quite well and knows what you are trying to achieve in your succession plans. While I would never expect this professional to be very close to your family, it would be hoped that they would know enough about your family to be able to make good judgments. Again, regular family meetings involving outsiders strengthens these arrangements.

Replacement trustees

The replacement trustees who will eventually replace you and your spouse are very important. You want someone who knows you very well – these people are going to have to exercise the discretion that the trust allows and make decisions for you. They will have your letter of wishes as guidance. However, you need someone who can confidently predict what you would have wanted in just about any circumstance. For example, your letter of wishes might say that you would like all three of your children to benefit equally. However, the accident that killed you and your spouse also severely disabled one of your children who will now need additional care for the rest of her life. The replacement trustees would put aside the letter of wishes (there are now circumstances that were never contemplated by you when you wrote it) and consider what they should do. In doing so, they would think about what you would have done in the same circumstance – trying to stand in your shoes.

These people therefore need to know you and your family very well. There are some things to think about when considering your replacement trustees:

- Do not have one of your children (or other beneficiaries) as a replacement trustee. This could be a recipe for disaster within your family. Remember that the replacement trustee

will decide what distributions will be made, and if it is one of the beneficiaries it could seem unfair to the others.
- A family member who is not a beneficiary (sibling, cousin, etc.) or a very close family friend can make a good replacement trustee. These are people who are close enough to you to be able to judge what you would want to happen in any particular circumstance, but distant enough to be impartial.
- Both you and your spouse should choose a replacement trustee. This protects each of you if the other remarries.
- The replacement trustee does not necessarily need a farming or business background. Remember that there will be (or ought to be!) a professional who can provide support and advice on these areas. The main task for your replacement trustee is to think and make decisions in much the same way as you would have done.
- The replacement trustee should not be one of the children's guardians. The fact that the guardian is required to approach the trust for funds for the children means that there are checks and balances in the system.
- The replacement trustee may be one of the executors of your estate.

The replacement trustee(s) will be named in your will and in many cases the appointor and the replacement trustee will be the same person. The identity of this person is something to which you should give a lot of thought.

Letter of wishes

This is also often known as a 'memorandum of wishes'. Whatever it is called, it is an informal document that effectively sets out what you would like to happen to the trust and its assets after you are gone or no longer able to act as a trustee. It cannot contradict the trust deed, nor does it bind the trustees. However, although the letter of wishes is not binding, given that the trust is discretionary in nature it is important to provide one. You should have a letter of wishes at all times and be sure that it reflects your current wishes. Although the trustees have complete discretion regarding which beneficiaries will benefit under the trust and are not obligated to follow your letter of wishes, it is extremely rare for trustees to disregard a settlor's wishes, especially if they have been put in writing. Moreover, it is unlikely that a replacement trustee would disregard your wishes for no good reason — the chances are that if the replacement trustee does not follow what you have specified it is for a good reason that is probably based on a change in circumstances.

There is no requirement for your wishes to be in writing — they are just as valid if you have verbally told your replacement trustees what you want to happen. However, for the sake of clarity, it is best to put your wishes on paper. Although your wishes are not really a legal matter, it is a good idea to work with your lawyer on this. Your lawyer should make sure that you have covered all areas in which there could be uncertainty and ensure that the letter of wishes is written in clear and unambiguous language. While this is an informal document, and most likely to be read mainly by the replacement trustee who knew you very well, it is possible that it could be read and interpreted by someone who has never known you (e.g. if your replacement trustee dies soon after you).

The letter of wishes gives you the opportunity to tell the replacement trustees what you want to happen. It is also an opportunity for you to clarify your thinking — often having to write something down means that you check what it is that you really want and also fill in the detail. Writing the letter of wishes may give some form to the discussions that you and your spouse need to have regarding succession issues. You should also review your letter of wishes quite regularly. Because it is only a statement of what you would like to see happen, you can update it at any time, i.e. when circumstances change or something happens to make you change your mind.

Some of the things that need to be covered are:

- *The extent to which the children will share equally.* While many non-farmers demand that in normal circumstances all children will share the trust fund equally, this is often not so in farming families. You no doubt have some idea of what you think is fair for the children – this should be set out in the letter of wishes.
- *The length of time for which the farm should remain in the trust.* It may be that you wish for the farm to stay in the trust for some years after you have gone, and that the child who wants to carry on farming will lease the land and buildings from the trust. Over the years, this child can gradually buy out his or her siblings' interest in the trust.
- *When to distribute.* The circumstances in which children should receive distributions, e.g. for education, health issues, home purchase, business start-up, etc.
- *When children will receive their major distributions.* This could be a specific age (e.g. thirty), though I would generally leave this to the judgment of the replacement trustees. The letter of wishes may give some guidance regarding this, e.g. that the children are standing on their own two feet professionally, have married or have children.
- *Whether the replacement trustees should talk to the children about their relationships before making major distributions.* At the very least, the replacement trustees should be sure that the children know the implications of the Property (Relationships) Act and how it would apply to their situation.

- *Making loans.* Whether the replacement trustees, if concerned about the relationships of the children, should consider making loans (effectively interest free) instead of gifting funds.

The letter of wishes will not control the family trust after you have gone, but it will guide it. That is how it should be: you do not want to bind the replacement trustees because they can then be prevented from responding to some unforeseen circumstance. You should make sure that you keep your letter of wishes up to date – does it reflect what you want to happen? Furthermore, keep telling your replacement trustees what you are thinking in terms of succession. Given that your replacement trustees are likely to be close family friends or relatives, you will see them often enough. Bring the conversation around to succession at times so that your replacement trustees will always be in a position to know what you would have wanted in any particular circumstance. Again, this is a good reason to involve these people in at least some of your family meetings.

CHAPTER 14

Being prescriptive

Your will

While the family trust is discretionary, your will is prescriptive. Through your will, there are many things that you can do that you cannot insist on through your trust. While there may be things that you request via your letter of wishes to the trustees of the family trust, these are not binding. Therefore, outcomes that you are certain you want to achieve, or particular items or assets that you wish to go to a particular person, must be organised through your will.

For those following a discretionary approach to succession arrangements, the will becomes less important: most of your significant assets will be dealt with by the family trust rather than your will. Nevertheless, your will does deal with some very important matters, in particular the nomination of guardians for your children (assuming that they are of an age where this is needed) and someone to become the appointor of the family trust and the replacement trustee. Because the farm and other important assets are owned by the family trust, you cannot dictate what will happen to them via you will. Your will can only deal with personal assets. While you may request what should happen to the trust's assets, and record your wishes, ultimately that will be up to the replacement trustees you appoint.

Some of the main provisions of your will are likely to be:

- all your major assets to go to your family trust;
- any debt owed by the trust to you to be forgiven;
- guardians for the children to be nominated;
- any specific assets that you want to go to particular people to be outlined. Note that these assets must be owned by you — your will cannot deal with assets that are owned by the trust;
- any specific requests (e.g. burial or cremation) to be made;
- appointor and replacement trustee for the family trust to be nominated. This will often be the same person who is the trustee/executor of your estate, but does not have to be if that is not suitable.

Although your will is no longer as important for dealing with your most valuable assets, it does need to be kept up to date. In particular you need to keep checking that the appointor/replacement trustee you have nominated is still the appropriate person. It is possible that the person whom you have nominated may become incapacitated, move out of New Zealand or die — you would then have to nominate another. It may no longer deal with your most significant assets, but your will is still a very important document and needs to be continually reviewed.

There does come a time when some farmers can be quite definite about what they want to happen. When you reach this point, it is time to have a major restructure — a new will and letter of wishes will have to be made and possibly some assets distributed out of the family trust. This will only happen when you are very sure of the path that succession will take. While the children are relatively young, and their path in life unsure, you will want to keep your options open. It is when your children (and you) are certain that they either want to farm or not to farm that you can take a prescriptive approach. This will follow a lot of discussion within your family and the family meetings described earlier. If you can get complete agreement within the family, you can restructure things and document them in such a way as to ensure they happen.

This may mean that the family trust is no longer the best ownership vehicle for some or all of your assets. The trouble with family trusts is that you cannot insist that something happens or that certain children get particular assets. This is fine and proper when everyone is younger and no definite plans are possible. However, it is not fine if you and the rest of the family are quite sure that you know what you want to happen with the farm and other assets. This may not happen until the children are all in their thirties or it may happen at the time when you are thinking of leaving the farm.

There is a range of options which are discussed here. However, it needs to be said at this stage that in the event that none of your children wants to take over the farm succession is very simple: you sell the farm, invest the proceeds and, when you die, the children each take an equal share of your estate or family trust. It is possible that you will help the children in some way or other (fund a business, help with the purchase of a house, contribute to the cost of a PhD, etc.) and that this financial help will be taken into account in the final wash up (i.e. when you have died). If none of your children wants to take over the farm, your succession issues are really no different from non-farming families.

However, if one (or more) of your children does want to take over the farm, there are four options:

1 *Just say no!*
This is not a particularly attractive option for most farming families – refusing to hand the farm down to the next generation and selling it to outsiders means that the farm moves out of the family. However the farm and other assets that you own are yours and you need something to fund your lifestyle after you have moved off the farm. Moreover, for some people (admittedly not large numbers) being able to pass wealth down to the children equally may be more important than keeping the farm in the family. When one of the children is quite clear about wanting to take over the farm, most farming families do whatever it takes to make this happen. Sometimes, however, that's too high a price – you should at least consider selling the farm and helping the child who wants to farm onto a farm in some other way.

2 *Distribute the farm to the child*
This means effectively gifting the farm, and being able to do this implies that you have other quite significant assets. You will need these other assets: partly to buy a house and have investment capital on which to live, and partly, perhaps, to be able to help the other non-farming children – at least after you are gone, if not in your lifetime. These demands on other assets mean that this option does not work for a lot of farmers. This is a great shame because this option means that the child who does want to take over the farm does not have to take on a lot of debt. Farmers who build up a significant portfolio of off-farm investments do have this as an option – off-farm investments are a good idea for a number of reasons and succession is one of them.

If this is to happen quite late in your life, it may be a good idea to transfer the farm out of the trust and into your own name. This means that you can deal with the farm under your will and have much more certainty about what will happen. Transferring the farm to your own name is most easily done if the farm is owned by a company and the shares of the company are owned by a family trust.

A variation on this would be for the family trust that owns the farm to distribute a part of the farm to the child. This may be done by distributing one or more of the titles (if the farm is in more than one title); but a more practicable way is to have the farm owned by a company, the shares of which are owned by the family trust. In this way you can distribute some shares, representing a part of the farm, and distribute more in the future as and when it is appropriate.

3 *Sell the farm to the child*

This option has the very positive benefit that you get some cash with which to buy a house and invest for income, and you have funds that you can leave to other children. The unattractive aspect is that your child is likely to have to borrow to fund the purchase (assuming the child has not won Lotto or robbed the local bank).

If the farm is owned by a company, you can also sell a part of the farm. This would be further complicated by the need to mortgage the farm in order for your child to buy into it, which would in turn put you at risk.

4 *Lease the farm to the child*

This would probably be achieved by the family trust leasing the farm to the child (or child's company or trust) who would then farm it as a business. The family trust would receive a rental that could be distributed to all the children if you didn't need it as income. The child who is farming would know that the share would come eventually and could gradually buy out siblings' interest in the farm by buying shares in the farm from the trust. Again, this would be complicated somewhat by the trust having to mortgage the farm in order for the farming child to be able to borrow to buy out the siblings' interest, and would leave the farming child's share in the farm exposed to a potential major farming downturn.

Chapter 15

Getting good advice

Advice is just that – advice. Just as you wouldn't let your farm adviser or accountant make all the decisions about your farming business neither should you turn over the management of your succession or investments completely to a financial adviser or broker. It's your money and your life and you should retain responsibility and control. The least effective wealth builders are those who turn over the *decisions* about their money to others. You must make these important decisions but there is plenty of help available.

Why use advisers?

Good advice in any field is invaluable. The more you know the faster you can go in any area whether it is in the development of your farm or growing wealth in off-farm investments. Knowledge is a form of wealth and it's usually worth the price. You are in charge of your succession strategies but you don't have to do it all alone. No matter how well you educate yourself or how well organised you are you will still need expert help from time to time. Experts have seen it all before and, very importantly, know what mistakes to avoid. If you are committed to getting to your last fencepost in the manner of your choice then you will make sure that you surround yourself with a 'dream team' of professionals who will assist you in making good decisions along the way and will provide good advice.

We all need some help at times and many of us rely on friends and family to support us. However, when you are dealing with your succession processes I think that you need to look to professionally qualified advisers. Unqualified friends and relatives, no matter how supportive, will not know enough to give you good advice about succession. And no matter how hard you work at learning about investments, trusts, tax or people development, you will never know everything and will also find it difficult to remain up to date in certain areas. Advisers are useful both in the good times and the bad – and if the going gets tough you certainly want to be able to call in a 'heavyweight' in the right area! Advisers are specialists too – just in a different field from you.

You are likely to need one or more of the following:

- lawyer;
- accountant;
- financial adviser;
- sharebroker;
- farm adviser;
- insurance broker.

It is never easy to choose professionals. It can be especially intimidating if you have seldom dealt with these kinds of people before. However, it's all common sense – just as when choosing a dentist or medical professional you need to ask around. No sensible farmer is going to pick key professionals from the Yellow Pages.

Neither does a sensible farmer want to meet any of these people for the first time when in trouble. You should start as early as possible to choose professionals and begin to build relationships with them so that they know you, where you want to be when you have put in the last fencepost, your family and your farming business and you are comfortable working with them.

Who you need to approach is determined by your circumstances and what you are planning to do for your succession. The more complicated your affairs and greater your goals, the more expert advice you will need. If for example you have share investments and earn some dividends a general accountant will be able to help you file your tax return and minimise your tax liabilities. However, if you are investing in several other farms, trading in shares or own a portfolio of rental properties you may need an accountant with more specialised knowledge in your area of investment activity.

Finding good professionals

The best way to find good professionals is to ask around; people who are considering succession issues themselves or who are active in the same investments as you will be especially helpful. Other property investors, for example, should be able to direct you to good real estate agents, valuers, mortgage brokers and accountants who specialise in property investment and the applicable tax law. Again, this is where it can be very useful to join investment groups, and network with people who are interested in the same investments as you and can recommend reliable professionals. However, the ultimate responsibility for choosing professionals rests with you. There are two main rules to keep in mind:

1. *Good professionals do what's right for you – not what's right for them.* Naturally, your advisers are in business and have to earn a living but it is you – as the client – who should come first.

That may mean that they have to advise you to do nothing or send your business away or recommend someone else – even though that means they may earn little or nothing. It can also mean that they may have to tell you at times that you are wrong – and risk incurring your wrath or losing your business. Such a professional is a very valuable adviser and is to be treasured.

2. *Good professionals advise you rather than sell to you.* One of the problems for many professionals especially in the area of finance is that they only get paid if you buy, sell or change something because their pay is based on commission or they get a fee for the transaction. This is why it is very important for you to understand how your adviser is paid. So it takes a very professional sharebroker to tell you to wait for a month or two or advise you to pay down debt instead.

So how do you find your key professionals? Well, obviously, they need to have the appropriate qualifications and experience – and no criminal convictions or bankruptcies. It helps to deal with established and reputable firms – you can expect that there is some supervision and systems in place for keeping the professionals up to date and complying with best practice. Word of mouth is probably still the best way to find the individuals that you can work with best so ask around. Ask particularly about levels of service and how satisfied your contacts are with how they are treated. When you have a shortlist of candidates you should meet with each one and ask lots of questions to establish if this professional is right for you. Questions might include:

- What are your qualifications?
- What experience do you have in working with people in my situation?
- What is a typical client like?

- What sorts of services do you provide?
- How do you work with your clients?
- How are complaints dealt with?
- What research and support do you have?
- What professional associations do you belong to?
- Does your association have a code of professional conduct?
- Are you affiliated with any other institutions? Are you independent?
- (Very importantly) how do you get paid?

He who pays the piper...

You need to know whether the professionals are paid by fee or commission. If your advisers are paid by commission it is in their best interests to sell you something – otherwise they do not get paid. Whoever writes the cheque is the employer to whom loyalty is owed! On the other hand, if your professionals are paid by fee it will cost you no matter what you decide to do but your adviser has no reason to steer you in the direction of anything that is not right for you so there is no conflict of interests. There is really no right answer in terms of how your advisers are paid but it is best that you know how they are remunerated before you begin the relationship.

Good advice can be expensive: lawyers and accountants will charge for their time; agents will often take a percentage fee; and financial planners, fund managers and risk advisers are often paid a commission based on what you purchase. One way or another, *you* will pay. That is why it is important that you understand how your advisers are paid – especially if it may impact in any way on the advice they offer. Remember, too, that many professionals will waive fees for initial meetings or attend some meetings for a reduced price. Negotiate! You can – and should – ask for estimates and quotes. However, I do not begrudge professionals their fees –

ultimately you get what you pay for. And it may cost you even more if you receive poor advice or seek no expert opinion at all.

In the final analysis, you will have to trust your own judgment regarding the person as almost all professionals will be prepared to give appropriate answers at your interview – that's why personal recommendations from contacts you trust are so valuable. You need to be comfortable with the people you choose as you will have to work closely with them and will also need to talk to them about many private matters – your finances, plans, hopes for the future. You are expecting to have a long-term – even a lifetime – relationship with these advisers so you should take your time and choose with care. But always remember that the adviser is working for you and that you are paying for the service. If you are not satisfied at any time, get a new adviser. Your investments and your future financial independence are far too important to be compromised by an incompetent or otherwise inadequate adviser.

Create a winning team

Winners work with winners – once you find one good professional they will lead you to others. Good people like to work with other good people and a good professional in any field can lead you to the others that you need. Ideally, you want to have a team of professionals who will work well together to ensure that your investment strategy is successful and will give you what you want. As the diagram shows you need each of your advisers to focus on what is right for you and your investments and you also need them to work together to help you achieve your goals.

FIGURE 3: A winning team

Getting what you want

Get the most from your advisers by being prepared and thinking through what you want from each of them. You will save lots of time (and money) by having all your information to hand and being organised for each meeting. In addition you should be clear about your brief to the adviser:

- What do I want to achieve?
- What advice or help do I need from this adviser?
- Which aspects of my affairs will I look after myself and which do I want to delegate?
- What reports and information do I want to receive?
- How much time will I have to spend in this area and what do I want to be actively involved in?
- What authority will I give my adviser to act and which decisions do I still want to make?

Take a sharebroker for instance. Depending on your skill level, the amount of time you are willing to devote to your portfolio and the level of interest you have in investment, you may make various choices about a broker. You may choose to do all your own buying and selling online at minimal cost because you watch the market intently and research what is happening in your areas of interest. On the other hand, without a broker relationship you will be unable to access many new listings or investments to which the brokers have access. You will also miss out on the research reports and advice that a broker can offer. You may have neither the skill to be nor the interest in being actively involved in your portfolio at all. In that case you might give your broker authority to buy and sell on your behalf, operating more like a fund manager than solely as a broker.

Many of the brokerage firms can also hold custody of your shares – allowing all of your dividends to be consolidated in one statement rather than arriving in dribs and drabs. This makes managing your tax returns much simpler and also takes away all the week-to-week banking, filing and administration tasks. You could choose to have all your dividends reinvested as the broker sees fit or choose to take them as income – how you choose to use and brief the adviser is your decision.

Likewise with a financial adviser: you need to decide what you want. Don't turn over control of your investments to an adviser – you wouldn't let a farm adviser run your farm. On the positive side a good financial adviser will be invaluable in helping you think through your approach and allocate your investment money to a range of investments that are appropriate for your profile. Without that kind of support you are likely to be too risk averse or take on too much risk for your circumstances.

As with every other aspect of your affairs, and succession in particular, you can delegate much of the work and decision-making but the ultimate responsibility is yours – to choose good advisers, give them an appropriate brief and manage them well. Smart farmers will not be intimidated by their advisers and will ensure that they get the best from each one. Don't be reactive – ask for what you want, and manage your advisory team to ensure you get what you need.

The pay-off for you

Advice costs, but it's worth it. The only thing that's more expensive than good advice is poor advice. If you make the effort to find good people and take the time to get them working well together you should be confident that over the longer term you will:

- create more wealth;
- get higher returns with less risk;
- worry less about your farm and your investments;
- work to sound principles and strategies;
- receive the support to stay the course;
- be more likely to reach your ultimate goals;
- have a smooth path towards – and ultimately successful – succession;
- be freed up to deal with other life priorities.

A special word of warning!

Farmers often have significant lump sums to invest. This could be because a large cheque has arrived at a certain time of year or because the farm has been sold. While you may have done a lot of thinking and planning regarding asset allocation and your investment strategy, beware of handing over large sums for investment. I have seen a number of clients who invested significant sums on the wrong day or at the wrong time and who have taken years to regain the initial value of their investment. This was not because the advice was poor or the asset allocation was wrong, but because all the money was invested at one time rather than drip fed into the market (see Dollar Cost Average, p. 96). Advisers who earn a commission on investments are incentivised to place the investment as soon as possible. This can have unfortunate consequences for the investor!

Chapter 16

After your last fencepost

I've dealt with many clients who are facing the transition from full-time business or farming to some form of 'retirement'. It isn't easy and it always throws up a lot of challenges and choices that are not only about money.

Full retirement or any level of scaling back from full responsibility for the farm involves a lot of lifestyle changes. There will be several questions you'll want to discuss together and be able to answer:

- What will you tell your family and friends?
- What will your new 'label' be?
- How will you spend your time?
- How will your relationship change?

There are several things you can consider, debate and decide in advance.

Redefining 'retirement'

'Retirement' is an unfortunate word in many ways. First it means 'to withdraw' which may be exactly the opposite of what you want. No wonder so many people resist the idea! It has all of the connotations of being put out to pasture. Secondly we are deterred by a hangover from our parents' generation, when retirees stopped. They got the gold watch or the send-off party and then the pipe, slippers and Lazyboy. Many had the newspapers, the garden and the occasional hobby to fill in the years between being shut out of the workplace and shuffling off this mortal coil. Not that it didn't suit some: depending on what they had been doing, many were ready to stop. After all this was a generation that had seen at least one world war, if not two. And many had worked physically very hard (women included) all of their lives and were weary. This generation is different. I try to avoid using the term 'retirement' as much as possible though it's almost unavoidable. But no matter what age or stage you and your spouse are at, you too are 'succeeding' to a new phase. Whether you and your partner intend to continue to work a lot, a little or not at all, you can be sure this period of your life is likely to be very different from that stage of life for previous generations. Have it on your own terms!

Age with attitude

Your age is only a number – and one of the least important in terms of how you choose to live. Ask yourself how 'old' you would be if you didn't know your own age. Put the effort into thinking and talking about what you want to do and who you want to be and, as much as you can afford to, live life on your own terms.

Don't make life small!

Life expectancy is elastic. Already men can expect to live to their late seventies and women to their early eighties – and these averages include all of the people who died young from disease or accidents. Every year you live increases the chances that you will live to be very old.

Remember that this is also the youngest generation of older people ever! You will have heard people say that fifty is the new forty and sixty is the new fifty and so on. While there may be an element of wishful thinking amongst we baby boomers, as we have always enjoyed occupying centre stage, there is nevertheless some truth in this. We had good nutrition, good healthcare and little in the way of armed conflict and have also benefited from massive advances in medical knowledge and technology. Our lives have been machine assisted as never before and we have not normally done the back-breaking work of earlier generations. Even if you just consider our farm bikes, tractor cabs and computers, you can see the very big differences in the way we have worked and lived. When we have had problems we have also had remedies – hip replacements, knee reconstructions, Voltaren and Valium! Even being wet and cold is at a new level of comfort with Gore-Tex and other advanced textiles.

All this means that when we hit midlife and wish to scale back our responsibilities we are by no means ready for the rocking chair. We are in great shape mentally and physically and have lots of options and energy for the second half. So ask yourself and your partner what game do you want to play and how do you want to play it?

Design a lifestyle

It can be very hard to know or decide what you will want a decade or two in the future but I do think you should make plans for the immediate months and years after putting in the last fencepost. A practical way to do this is to get some annual time planner charts and start to block out parts of year one – perhaps with travel, a conference in an area of interest or an aspect of investment, some major family events you want to attend and a course you'd like to take.

Similarly, it is very useful to use a chart to plan a typical month: you might include a weekend away, your meeting or sporting commitments, a movie, time for visiting friends and time for the grandchildren.

Think also in terms of typical weeks – both winter and summer. This kind of exercise can help you give shape to your new lives. Your previous annual cycles on the farm were probably a given, as you had to do what had to be done with lambing, calving or cropping – or all three! You may never before have had the luxury of choice. Too much choice can be paralysing however and many clients have found it very useful to put some structure into the early days of this period of change. After a while, the rhythm of the things you commit to and want to do will soon take over, but a plan helps take away the initial feelings of unease.

'Rework' the idea of retirement

Work gets a bad rap! We have all tended to assume that work is stuff you wouldn't do if you didn't have to. Many people think they only work to put food on the table and pay the bills. However, I beg to differ. Work is really about any effort expended to meet a goal. We 'work' when we walk (for pleasure) up a hill, hit golf balls on a wonderful autumn day in Central Otago, cast our line into a Southland stream, bake bread or read stories to grandchildren. Work matters. My point is that work does not have to be unpleasant. And remember that all compulsion is now gone; you are free to choose what you will do. Almost all endeavours have aspects we enjoy less: tidying up after a wonderful family meal; putting away the tools after a marathon gardening session; cleaning out the car after a fishing trip. It's the same with business activities; various people may hate devising the agenda, attending the meeting, writing the minutes or doing the follow up. You may enjoy many aspects of managing your wealth – but, for example, hate the filing and administration. However, it is highly unlikely that you wish to do nothing. Work provides a great deal more than income: important things like structure, challenge, a sense of self worth and social interaction. Depending on your age and energy it is also likely that you still want some involvement in farming, business or other income-earning activities.

Farmers have lots of choices. After all, you have run a business for decades and are an expert in the country's primary export sector. And don't forget:

- there is a labour shortage and will be for the foreseeable future;
- the world is seeking knowledge workers and people who are business literate;
- you have knowledge and skills you can apply as an instructor, teacher, coach or mentor in several areas;
- you are likely to be wanted as an adviser or mentor for other farms or allied businesses.

Make your own rules

The rules have changed. It used to be that you went grey in middle age, bought clothes with ever-expanding waistlines (or no waist at all) and quietened down – lived more sedately, stayed at home, moved in a smaller circle, put your life at the service of the grandchildren. You can still choose any or all of these things if you wish but you also have lots of other options:

- different paid work;
- voluntary work;
- full-time hobbies;

- investment activities and management;
- physical activities;
- travel – for pleasure or learning, or with a purpose such as aid;
- consultancy – back to the farm or to another farm or business;
- directorships.

Assuming you have the means to be, do and have what you want, you have no barriers except in your own mind. Don't let anyone stop you from living the second stage as you wish. You don't need a role model: be your own. Shirley Conran was famous for saying in the 1970s that life was too short to stuff a mushroom! She was right. I'd go further: life's too short for anything that you don't really want to do. You have earned the right to let go of the 'shoulds' and 'oughts' and need now to get busy on the 'love to' and 'want to' lists. Why waste these wonderful years on anything else? Enjoy them on your terms. Nearly all the so-called 'rules' by which we live our lives are made up anyway – you can write your own! What are *you* doing after you have put in your last fencepost?

Leave a special legacy

There's a lot more to the concept of 'legacy' than mere money. Have a think about what your children and grandchildren might *really* need or want you to leave behind:

- a history of the family farm;
- a collection of farm memorabilia;
- photographs or artwork of the farm;
- family heirlooms that have been collected or restored;
- stories or recordings about the past.

These sorts of legacies reach beyond the grave in a very special and timeless way.

Second half – second chances

One of the great privileges of our generation is that we are still young and healthy enough in middle age to have a second chance in many areas. It's not too late to do that reading or formal study that you always wished you had done as a school leaver. You can resume those sporting interests that you abandoned because of the farm or family; in fact you can take up any or several of the new ones on offer, such as mountain biking, kayaking, triathlons, Pilates. This generation has a Peter Pan muse; we can do almost anything that we would have liked to have done in our twenties, the difference being that this time we can probably afford to pay for it! So think about the things that attract you now or that did back then and take that second chance to do what you want to do.

Don't be in a hurry to give money away to the next generation

You have a lot of living left to do, and many things may change over the coming years. I understand the urge to support children, and the joy of giving family members a hand to fulfil their dreams. However, I do not think that you should do this at the expense of your own options in the second half of your life. You need to keep your options open and live confidently for the future.

Life isn't static; neither should your succession plan be set in concrete. You must continually respond to events as they occur and review and amend your plan as necessary.

Dream come true

Successful succession is all about achieving your dreams for your farm and your family. When it works well it looks as if it were meant to be, and appears effortless and seamless. The reality is somewhat different: managing a successful succession means you have to plan, communicate, develop and invest to secure the outcomes you want. It takes time, effort and commitment.

There is also a difference between the people who get it right and those who get it wrong. The difference is a combination of attitude, communication skills, clarity of purpose and application. Again, the 'soft' stuff is really the hardest to do well; the 'hard' stuff, such as business structures, trusts and wills, follows logically. Some luck – with your spouse, your children, your farm – helps, but I wouldn't rely on it. You get the end result of the set of choices you make and the actions you take – or don't take – along the way.

Succession will happen some day anyway, with or without all this work. It just may not be the succession that you want. Inaction gets results as well; just not the desired results. The problem only gets bigger all the time.

It's never too early to begin; it's almost never too late.